Vespa Ciao and Bravo Owners Workshop Manual

by Pete Shoemark

Models covered:
All Ciao models, First introduced into UK October 1968
All Bravo models, First introduced into UK late 1977

ISBN 0 85696 374 7

© Haynes Publishing Group 1978

ABCDE
FGHIJ
KLMNO
PQRS

All rights reserved. No part of this book may be reproduced or transmitted in any form or by any means, electronic or mechanical, including photocopying, recording or by any information storage or retrieval system, without permission in writing from the copyright holder.

Printed in England

HAYNES PUBLISHING GROUP
SPARKFORD YEOVIL SOMERSET ENGLAND
distributed in the USA by
HAYNES PUBLICATIONS INC
861 LAWRENCE DRIVE
NEWBURY PARK
CALIFORNIA 91320
USA

Acknowledgements

Our thanks are due to Douglas (Sales and Service) Ltd of Bristol, and to Mrs D.I. Shoemark, who supplied respectively the Vespa Bravo and Ciao mopeds featured in this Manual.

Brian Horsfall assisted with the stripdown and rebuilding, and devised the ingenious methods for overcoming the lack of service tools. Les Brazier arranged and took the photographs which accompany the text, and the cover and inside cover photographs.

Jeff Clew edited the text and compiled the technical data used in the manual, in conjunction with Roy Stone, of Douglas (Sales and Service) Ltd.

Finally, we would like to thank the Avon Rubber Company, who kindly supplied information and technical assistance on tyre fitting, and NGK Spark Plugs (UK) Limited for information on sparking plug maintenance and electrode condition.

About this manual

The author of this manual has the conviction that the only way in which a meaningful and easy to follow text can be written is first to do the work himself, under conditions similar to those found in the average household. As a result, the hands seen in the photographs are those of the author. Even the machines are not new; examples that have covered a considerable mileage were selected, so that the conditions encountered would be typical of those found by the average owner/rider. Unless specially mentioned and therefore considered essential, Vespa service tools have not been used. There is invariably alternative means of loosening or slackening some vital component, when service tools are not available and risk of damage is to be avoided at all costs.

Each of the six Chapters is divided into numbered sections. Within the sections are numbered paragraphs. Cross-reference throughout this manual is quite straightforward and logical.

When reference is made, 'See Section 6.10', it means Section 6, paragraph 10 in the same Chapter. If another Chapter were meant it would say, 'See Chapter 2, Section 6.10'.

All photographs are captioned with the number of the Section/paragraph to which they refer, and are always relevant to the Chapter text adjacent.

Figure numbers (usually line illustrations) appear in numerical order, with a given Chapter. 'Fig. 1.1' therefore refers to the first figure in Chapter 1.

Left-hand and right-hand descriptions of the machines and their components refer to the left and right of a given machine when normally seated.

Whilst every care is taken to ensure that the information in this manual is correct no liability can be accepted by the authors or publishers for loss, damage or injury caused by any errors in or omissions from the information given.

Contents

Chapter	Section	Page	Section	Page
Introductory sections	Acknowledgements	2	General repair information	11
	About this manual	2	Quick glance maintenance,	
	Introduction to Vespa Ciao		adjustments and capacities	12
	and Bravo mopeds	6	Recommended lubricants	12
	Ordering spare parts	7	Working conditions and tools	13
	Routine maintenance	8–10		
1 Part 1: Engine	Operations: Engine in frame	15	Crankshaft assembly	18
	Removed from frame	15	Piston and rings	19
	Removing unit	15	Reassembly general	22
	Dismantling – general	16	Cylinder barrel – replacing	24
	Cylinder head, barrel		Refitting in frame	25
	and piston removal	17	Starting and running the	
	Crankcase halves	18	rebuilt unit	25
1 Part 2: Transmission	Clutch unit removal	28	Automatic speed governor	35
	Rear pulley – single speed	28	Automatic clutch unit	36
	Rear hub unit – single speed	30	Chainwheel, pedals and	
	Rear hub unit – variable ratio	32	pedal shafts	37
	Pedals, chain and freewheel unit	35	Drivebelt adjustment	37
2 Fuel system and lubrication	Petrol tank (Bravo only)	39	Carburettor	40–44
	Petrol tank flushing	39	Air cleaner	44
	Petrol tap	40	Exhaust system	44
3 Ignition system	Timing – checking	46	Coil – removal	47
	Contact breaker	46	High tension lead	48
	Condenser	47	Spark plug – checking	48
4 Frame and forks	Rigid front forks	49	Rear suspension units	53
	Leading link front forks	50	Centre stand	55
	Front forks	51–53	Saddle – adjustment	55
	Frame examination	53	Speedometer head	59
	Swinging arm rear suspension	53	Speedometer drive cable	59
			Cleaning the machine	59
5 Wheels, brakes and tyres	Front wheel	61	Rear wheel – removal	65
	Front brake	62	Rear wheel bearings	65
	Front wheel bearings	62	Adjusting front and rear	
	Speedometer drive gearbox	62	brakes	65
	Rear wheel – examination	62	Tyres	68
6 Electrical system	Checking the electrical		Tail lamp	71
	system – general	69	Horn location	71
	Headlamp	70	Wiring layout and	
	Horn and light switches	70	examination	71

Note: General description and specifications are given in each Chapter immediately after list of contents.
Fault diagnosis is given when applicable at the end of the Chapter.

Metric conversion tables 74–75
Index 76–78

Left-hand view of 1973 Vespa Ciao moped

Right-hand view of 1978 Vespa Bravo

Introduction to the Vespa Ciao and Bravo mopeds

The Vespa name, meaning 'wasp' in Italian, has long been associated with the distinctive scooters produced by Piaggio and C of Genoa. Developed soon after the end of the second World War, these machines first appeared in the UK in 1949, and for many years were manufactured here by Douglas (Sales and Service) Limited.

In 1964, production of the British made machines ceased, and Douglas turned their attention to the importation of the Italian models in knocked down form, final assembly taking place in their Bristol works. The popularity of Vespa scooters increased over the years, culminating in the scooter boom of the mid to late 1960's.

Dr. Piaggio's original concept of the Vespa scooter has proven sufficiently sound to warrant no more than refinement in over thirty years of production. Piaggio applied the principles of the Vespa scooters when designing the Ciao moped, which was first announced in 1967. The original E1 model featured a single speed fan-cooled two-stroke engine, leading link front forks and direct lighting. The V1 model was similar, but featured a variable final drive ratio, to give improved acceleration.

In 1971, the E1 and V1 models were discontinued, being replaced by the Ciao Super Comfort model. This was very similar to the E1, but was equipped with smaller 17 inch wheels and an improved saddle. March 1974 saw the introduction of a second model, the Super Comfort V, with variable speed transmission, the original model being known as the Super Comfort E.

1977 saw the introduction of the Bravo models, which retain most of the features of the Ciao, including the engine unit and the choice of two types of transmission. These models feature a tubular main frame member beneath which the engine and transmission assembly are pivoted, to provide a form of swinging arm rear suspension. Bravo models are available with two types of telescopic front suspension similar to that used on conventional motorcycles.

Dimensions and weights

Dimensions	Bravo	Ciao
Overall length	1590 mm (62·5 in)	1570 mm (61·8 in)
Overall width	640 mm (25·2 in)	630 mm (24·8 in)
Overall height	1030 mm (40·5 in)	995 mm (39·2)
Weight (dry)	37 – 42 kg (81 – 93 lbs)	33·5 – 37·5 kg (74 – 85·5 lbs)

Ordering spare parts

When ordering spare parts for the Ciao and Bravo, it is advisable to deal direct with an official Vespa agent, who will be able to supply many of the items required ex-stock. Although parts can be ordered from Douglas direct, it is preferable to route the order via a local agent even if the parts are not available from stock. He is in a better position to specify exactly the parts required and to identify the relevant spare part numbers so that there is less chance of the wrong part being supplied by the manufacturer due to a vague or incomplete description.

When ordering spares, always quote the frame and engine numbers in full, together with any prefixes or suffixes in the form of letters. The frame number is found stamped on the lower part of the frame. The engine number is stamped on the bottom face of the crankcase (See accompanying photographs).

Use only parts of genuine Vespa manufacture. A few pattern parts are available, sometimes at cheaper prices, but there is no guarantee that they will give such good service as the originals they replace. Retain any worn or broken parts until the replacements have been obtained; they are sometimes needed as a pattern to help identify the correct replacement when design changes have been made during a production run.

Some of the more expendable parts such as spark plugs, bulbs, tyres, oils and greases etc., can be obtained from accessory shops and motor factors, who have convenient opening hours, charge lower prices and can often be found not far from home. It is also possible to obtain parts on a Mail Order basis from a number of specialists who advertise regularly in the motor cycle magazines.

Engine number location (Bravo)

Frame number location (Bravo)

Routine maintenance

Periodic routine maintenance is a continuous process that begins immediately the machine is used. It must be carried out at specified mileage recordings or on a calendar basis if the machine is not used frequently, whichever is the sooner. Maintenance should always be regarded as an insurance policy, to help keep the machine in the peak of condition and to ensure long, trouble-free service. It has the additional benefit of giving early warning of any faults that may develop and will act as a safety check, to the obvious benefit of both rider and machine alike.

The various maintenance tasks are described under their respective mileage and calendar headings. Accompanying diagrams are provided, where necessary. It should be remembered that the interval between the various maintenance tasks serves only as a guide. As the machine gets older or is used under particularly adverse conditions, it is advisable to reduce the amount between each check.

Some of the tasks are described in detail, where they are not mentioned fully in the text. If a specific item is mentioned but not described in detail, it will be covered fully in the appropriate Chapter.

No special tools are required for the normal routine maintenance tasks. The tools contained in the toolkit supplied with every new machine will prove adequate for each task, but if they are not available, the tools found in the average household should suffice.

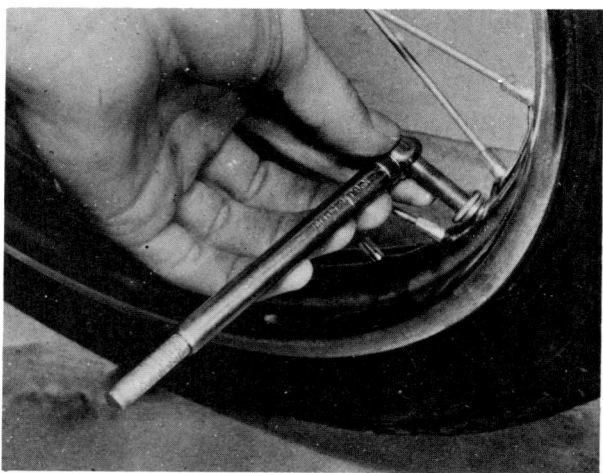

Check tyre pressures using an accurate gauge

Weekly, or every 200 miles (300 km)

1 Check the tyre pressures. Always check with the tyres cold, using a pressure gauge known to be accurate. It is recommended that a pocket pressure gauge is purchased to offset any fluctuation between garage forecourt instruments. On models fitted with 19 in wheels, the pressures are as follows:

Ciao models:
Front   20 psi (1.4 kg/cm^2)
Rear   35.5 psi (2.5 kg/cm^2)

On later machines, where 17 in wheels are employed, the tyre pressures should be as follows:

Bravo models:
Front   17 psi (1.2 kg/cm^2)
Rear   28 psi (2.0 kg/cm^2)

2 Legal check

Check the operation of the electrical system, ensuring that the lights and horn are working properly and that the lenses are clean. Note that in the UK it is an offence to use a vehicle on which the lights are defective. This applies even when the machine is used during daylight hours. The horn is also a statutory requirement.

Give each tyre a quick visual check for cuts or splits, and check that there is a reasonable amount of tread left.

Monthly, or every 800 miles (1300 km)

Complete all the checks listed in the weekly/200 mile service, and then the following additional items:

1 Sparking plug:

Remove the sparking plug, using a proper sparking plug spanner, to avoid any risk of damage to the ceramic insulator. Examine the colour and condition of the electrodes, comparing this with the sparking plug condition chart in Chapter 2. This will give an indication of the general condition of the engine. Clean the plug electrodes using a wire brush and a small magneto file or fine emery cloth. If the outer electrode is thin, or the centre electrode has been eroded excessively, the plug must be renewed. The gap can be measured with a feeler gauge, and should be 0·5 mm (0·020 in).

If necessary, adjust the gap by bending the outer electrode. On no account should any attempt be made to bend the inner electrode, or damage to the ceramic insulator nose will almost certainly result. Clean the sparking plug threads and wipe them with a trace of graphited grease. Refit the plug by hand, then tighten it carefully with the plug spanner, without overtightening.

Six monthly, or every 2500 miles (4000 km)

Complete all the checks listed in the weekly/200 mile and monthly/800 mile service, and then the following additional items:

1 Decarbonisation

Detach the decompressor cable at the cylinder head, slacken the three cylinder head nuts and lift the cylinder head away from its retaining studs.

Remove any accumulations of carbon from the combustion chamber in the cylinder head, and from the piston crown, using a blunt scraper. Ensure that the aluminium alloy parts are not damaged during this process. If desired, the combustion chamber may be finally polished using a proprietary metal polish. This will lessen the tendency for carbon to accumulate in the future. Before reassembling, ensure that all traces of carbon and metal polish are removed, paying particular attention to the sparking plug threads in the cylinder head.

It is also important to remove any accumulations of carbon in the exhaust port after detaching the exhaust system. If this area is ignored it is possible that the port will become severely restricted and that engine performance will become very much impaired. The carbon deposits in this area tend to be very hard, and it will probably be necessary to use a screwdriver blade or similar to chip them away. It is also possible to obtain specially designed wire brushes for use in electric drills. These can be used to good effect in the exhaust port. Note that decarbonisation is covered in detail in Chapter 1, Sections 11 and 13, of this Manual.

2 Cleaning the exhaust system

The exhaust pipe and silencer take the form of a welded system on both Ciao and Bravo mopeds, and the two cannot therefore be separated for cleaning. The normal soft carbon deposits which accumulate during use, form mainly in the short tailpipe at the rear of the silencer. This can be removed by way of a length of stiff wire passed up the tailpipe, and should be done each time the engine is decarbonised. Should the silencer appear badly choked, a condition which will produce a drastic drop in power, the silencer assembly should be detached from the machine and cleaned out carefully as described in Chapter 2, Section 10.

3 Adjusting the drive belt tension (single speed models only)

It is important that the correct belt tension is maintained on single speed models. If the belt becomes too slack, slipping will result, causing a noticeable power loss in use. It is equally important, however, that the belt is not overtightened as this will place undue strain on the engine main bearings and rear hub.

Start by releasing the left-hand outer cover to expose the belt and pulleys. Measure along the top run of the belt a distance of 315 mm (12·4 in), from the rear pulley. Using either weights or a spring balance attached at this point, apply a downward loading of 2 kg (4·4 lbs). This should cause the belt to be deflected by 13·5 – 15·5 mm (0·5 – 0·6 in). If the belt deflection is outside this figure, it should be re-tensioned as follows:

Referring to the accompanying photograph, slacken the engine mounting bolts (marked A). The silencer mounting bolt should also be slackened to allow it to slide against the frame. The engine can now be moved backwards or forwards by way of the small lever shown, to obtain the correct belt tension. (Note that the position of the adjusting lever varies, depending on the model).

Tighten the engine mounting and silencer retaining bolts, and refit the left-hand outer cover.

4 Carburettor adjustment

Carburettor adjustment should be carried out with the machine on its centre stand and the engine at normal operating temperature. A throttle stop screw is provided and may be reached by passing a screwdriver through the hole in the right-hand engine plate, having first detached the right-hand outer cover.

Start the engine and note the tickover speed. The speed can be reduced by turning the screw anti-clockwise and vice-versa. Try to obtain a slow, even tickover, and check that the engine does not falter when the throttle is opened quickly. When the tickover speed is set correctly, remove the top engine cover after releasing the single screw which retains it. Check the amount of free play in the throttle cable where it enters the right-angled cable guide. There should be about 2 – 3 mm ($\frac{1}{8}$ in) free play with the throttle closed. The cable guide is provided with an adjuster and locknut, to enable the correct amount of free play to be maintained. Should further attention to the carburettor be required, reference should be made to Chapter 2, Sections 6 and 7.

Yearly, or every 4800 miles (8000 km)

Complete all the checks listed in the weekly/200 miles, monthly/800 mile and six monthly/2,500 mile services, and then the following additional items:

1 Contact breaker points – adjustment

Remove the left-hand outer cover, and turn the engine over until the rubber plug in the flywheel is visible through the gap in the engine plate. Prise out the plug, and rotate the engine slowly until the contact breaker points are fully opened. Check that the point faces are not excessively burnt or pitted – if they are, they should be removed for renovation or renewal as described in Chapter 3, Section 4. If the contacts appear to be in good condition, measure the gap, using a feeler gauge. A 0·4 mm (0·015 in) gauge should be a light sliding fit. If the contact breaker requires adjustment, slacken the securing screw **just** enough to permit the fixed contact assembly to be moved, using a small screwdriver. Tighten the securing screw and then recheck the gap setting. Do not omit the rubber plug, which is fitted to prevent the ingress of water or other contaminants.

2 Rear hub unit – checking the oil level

The rear hub unit is fitted with a combined filler/level plug adjacent to the button which disengages the engine drive to the rear wheel. Access to the plug is gained after the left-hand outer cover has been removed. Unscrew the plug and check that the oil level is just up to the bottom of the hole. Top up, if necessary, with SAE EP 90 gear oil.

3 Hydraulic front forks – checking the oil level

On models fitted with hydraulic front forks, remove the level plug from each lower fork leg and check that the oil level is up to the holes(s). If necessary, top up, using SAE 20W engine oil. The oil can be introduced into the forks by means of an oil can.

Slacken bolts marked A to adjust belt tension

Routine maintenance

Hub should be topped up to level hole with SAE EP90 oil

Note position of fork oil level plug (marked OLIO)

General lubrication

In addition to the specific lubrication tasks detailed in the preceding Sections, the following items should be attended to periodically. The exact intervals at which these are undertaken depend largely on the amount of use the machine is given, and the conditions under which it operates.

1 Control cable lubrication

Apply a few drops of oil to the top of each of the four control cables at frequent intervals (preferably weekly). This regular lubrication operation will supplement full lubrication, which should be carried out as follows:

Disconnect the top of the cable in question, and build up a small cap of plasticine or similar around the top of the outer cable. Lodge the cable in an upright position, and fill the cup with light machine oil or engine oil, and leave the oil to drain through, preferably overnight.

A quicker and more positive method of lubrication is to use an hydraulic cable oiler which is fairly inexpensive and can be obtained from many motorcycle shops or by mail order from companies advertising in the motorcycle press.

2 Leading link front suspension – lubrication

The leading link pivot bolts should be lubricated with grease approximately once each year. This operation necessitates the removal of the front suspension, and reference should be made to Chapter 4, Section 3. Later models are fitted with a grease nipple, which simplifies the operation considerably.

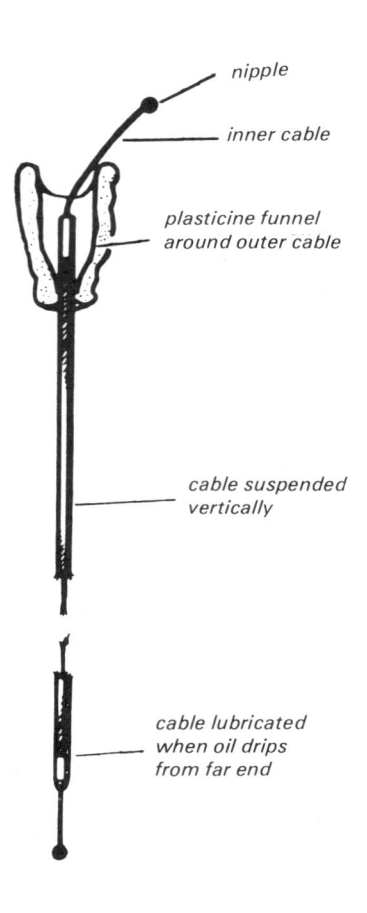

nipple
inner cable
plasticine funnel around outer cable
cable suspended vertically
cable lubricated when oil drips from far end

Cable ends are retained by pressed steel cleats

Control cable oiling

General repair information

In this Section, information is given to help answer some of the more common queries that may arise in the mind of the owner. Most of the answers are relatively simple with hints for simplifying the task, but where a major operation has to be undertaken, reference is made to the main text, thus alerting the owner that all may not be quite so simple and straightforward.

1 Repair of frayed cables

If, during the weekly maintenance session, any of the cables are found to be frayed with one or two strands of the inner cable broken they should be renewed.

Broken strands of cable can cause a cable to jam with disastrous results such as the throttle jammed wide open or the brakes jammed full on. Cables cannot be repaired successfully as once one or two of the cable strands have broken, the remainder will shortly follow. The relatively low cost of a new cable makes repair an uneconomical proposition.

2 Renewing the front and rear brake cables

Release the inner cable from the brake drum concerned by removing the small clamp which retains the cable to the actuating lever.

Disengage the cable from the handlebar lever, and pull the cable clear of the frame. Replacement is a direct reversal of the removal process, noting that the cable must be adjusted before the machine is used.

3 Renewing the decompressor cable

The decompressor cable nipple will unhook from the cylinder head arm and the cable will pull out of the cable stop. Pull the outer cable clear of the handlebar stop, turn the inner cable to form a right angle with the handlebar lever and pull the nipple out of the lever. The cable can then be pulled out of the cable guide on the steering head.

4 Renewing the throttle cable

The throttle cable is operated by a spiral cut in the throttle twistgrip sleeve. As this is turned, a block attached to the end of the cable is drawn along a groove in the twistgrip unit, providing a straight pull on the cable. The cable end is secured in the block by a small grub screw, and can be released after the unit has been removed from the handlebar end. It is clamped by a tapered screw and nut.

The lower end of the cable is retained inside the throttle valve, and can be disengaged after the top plastic engine cover has been removed, and the carburettor top released. A small barrel nipple on the cable end will be found to engage in a slot in the throttle valve, and can be displaced to release the cable.

5 Adjusting the handlebars – rigid and leading link fork models

Place the machine on its centre stand and slacken the chromium plated centre bolt at the top of the handlebar stem by about three turns. Tap the bolt head with a mallet to release the cone at the bottom of the stem. The handlebars can now be raised or lowered to the desired position and the bolt retightened to lock them. Ensure that the wheel and the handlebars are at 90° before final tightening.

6 Adjusting the handlebars – hydraulic fork models

This type of handlebar fitting precludes any height adjustment. They can, however, be set for reach by slackening the two U bolts which clamp the handlebars to the top fork yoke. Tighten the U bolts fully, after adjustment.

7 Renewing bulbs

Various types of headlamp and rear lamp units have been fitted to the Vespa Ciao and Bravo mopeds. In view of the different fixing methods involved it is advised that reference be made to Chapter 6, Section 3, for details of how access to the bulbs is gained. The headlamp bulb can be removed after the contact has been displaced sideways. The pilot and rear lamp bulbs are of the festoon type and are a push fit between two spring contacts.

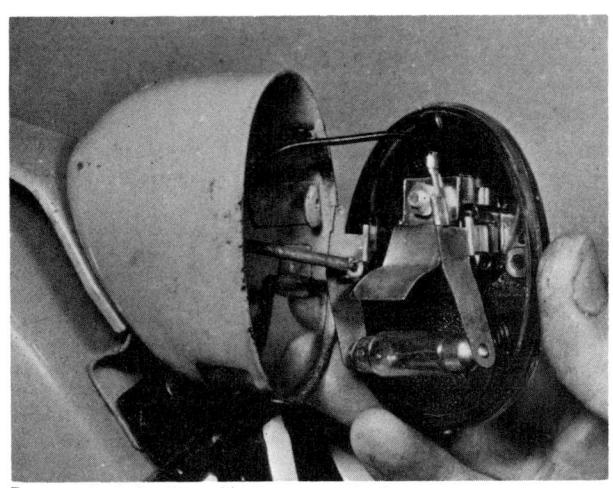

Displace contact to release headlamp bulb

Festoon bulb is clipped between contacts

Quick glance maintenance adjustments and capacities

Engine		SAE 30 self-mixing two-stroke oil at 2% or 50:1 ($\frac{1}{4}$ pint oil to 1$\frac{1}{2}$ gallons petrol), (20 cc to 1 litre petrol)
Rear hub unit		SAE EP90 gear oil. Fill to level hole
Front (hydraulic) forks		SAE 20W engine oil or fork oil. Fill to level hole
Contact breaker gap		0.4 mm (0.015 in)
Sparking plug gap		0.5 mm (0.020 in)
Tyre pressures		Ciao	Bravo
Front		20 psi (1.4 kg/cm^2)	17 psi (1.2 kg/cm^2)
Rear		35 psi (2.5 kg/cm^2)	28 psi (2 kg/cm^2)

Recommended lubricants

Component	Type of Lubricant
Engine	Self-mixing two-stroke oil. Mix with petrol at a ratio of 2% or $\frac{1}{4}$ pint oil per 1$\frac{1}{2}$ gallons of petrol
Rear hub unit	SAE EP90 gear oil
All greasing points	High melting point grease
Cables etc. (not nylon lined)	Light machine oil
Front forks (hydraulic)	SAE 20W engine oil

Working conditions and tools

When a major overhaul is contemplated, it is important that a clean, well-lit working space is available, equipped with a workbench and vice, and with space for laying out or storing the dismantled assemblies in an orderly manner where they are unlikely to be disturbed. The use of a good workshop will give the satisfaction of work done in comfort and without haste, where there is little chance of the machine being dismantled and reassembled in anything other than clean surroundings. Unfortunately, these ideal working conditions are not always practicable and under these latter circumstances when improvisation is called for, extra care and time will be needed.

The other essential requirement is a comprehensive set of good quality tools. Quality is of prime importance since cheap tools will prove expensive in the long run if they slip or break and damage the components to which they are applied. A good quality tool will last a long time, and more than justify the cost. The basis of any tool kit is a set of open-ended spanners, which can be used on almost any part of the machine to which there is reasonable access. A set of ring spanners makes a useful addition, since they can be used on nuts that are very tight or where access is restricted. Where the cost has to be kept within reasonable bounds, a compromise can be effected with a set of combination spanners — open-ended at one end and having a ring of the same size on the other end. Socket spanners may also be considered a good investment, a basic $\frac{3}{8}$ in or $\frac{1}{2}$ in drive kit comprising a ratchet handle and a small number of socket heads, if money is limited. Additional sockets can be purchased, as and when they are required. Provided they are slim in profile, sockets will reach nuts or bolts that are deeply recessed. When purchasing spanners of any kind, make sure the correct size standard is purchased. Almost all machines manufactured outside the UK and the USA have metric nuts and bolts, whilst those produced in Britain have BSF or BSW sizes. The standard used in the USA is AF, which is also found on some of the later British machines. Other tools that should be included in the kit are a range of crosshead screwdrivers, a pair of pliers and a hammer.

When considering the purchase of tools, it should be remembered that by carrying out the work oneself, a large proportion of the normal repair cost, made up by labour charges, will be saved. The economy made on even a minor overhaul will go a long way towards the improvement of a tool kit.

In addition to the basic tool kit, certain additional tools can prove invaluable when they are close to hand, to help speed up a multitude of repetitive jobs. For example, an impact screwdriver will ease the removal of screws that have been tightened by a similar tool, during assembly, without a risk of damaging the screw heads. And, of course, it can be used again to retighten the screws, to ensure an oil or airtight seal results. Circlip pliers have their uses too, since gear pinions, shafts and similar components are frequently retained by circlips that are not too easily displaced by a screwdriver. There are two types of circlip pliers, one for internal and one for external circlips. They may also have straight or right-angled jaws.

One of the most useful of all tools is the torque wrench, a form of spanner that can be adjusted to slip when a measured amount of force is applied to any bolt or nut. Torque wrench settings are given in almost every modern workshop or service manual, where the extent is given to which a complex component, such as a cylinder head, can be tightened without fear of distortion or leakage. The tightening of bearing caps is yet another example. Overtightening will stretch or even break bolts, necessitating extra work to extract the broken portions.

As may be expected, the more sophisticated the machine, the greater is the number of tools likely to be required if it is to be kept in first class condition by the home mechanic. Unfortunately there are certain jobs which cannot be accomplished successfully without the correct equipment and although there is invariably a specialist who will undertake the work for a fee, the home mechanic will have to dig more deeply in his pocket for the purchase of similar equipment if he does not wish to employ the services of others. Here a word of caution is necessary, since some of these jobs are best left to the expert. Although an electrical multimeter of the AVO type will prove helpful in tracing electrical faults, in inexperienced hands it may irrevocably damage some of the electrical components if a test current is passed through them in the wrong direction. This can apply to the synchronisation of twin or multiple carburettors too, where a certain amount of expertise is needed when setting them up with vacuum gauges. These are, however, exceptions. Some instruments, such as a strobe lamp, are virtually essential when checking the timing of a machine powered by a CDI ignition system. In short, do not purchase any of these special items unless you have the experience to use them correctly.

Although this manual shows how components can be removed and replaced without the use of special service tools (unless absolutely essential), it is worthwhile giving consideration to the purchase of the more commonly used tools if the machine is regarded as a long term purchase. Whilst the alternative methods suggested will remove and replace parts without risk of damage, the use of the special tools recommended and sold by the manufacturer will invariably save time.

Chapter 1 Part 1 Engine unit

Contents

General description ... 1	Cylinder barrel: examination and renovation ... 12
Operations with the engine in the frame ... 2	Cylinder head: examination and renovation ... 13
Operations with the engine removed from the frame ... 3	Engine reassembly: general ... 14
Removing the engine from the frame ... 4	Engine reassembly: joining the crankcase halves ... 15
Dismantling the engine: general ... 5	Engine reassembly: replacing the piston ... 16
Dismantling the engine: removing the cylinder head, barrel and piston ... 6	Engine reassembly: replacing the cylinder barrel ... 17
Dismantling the engine: removing the flywheel rotor ... 7	Engine reassembly: refitting the cylinder head ... 18
Dismantling the engine: separating the crankcase halves ... 8	Engine reassembly: refitting the flywheel rotor and fan ducting ... 19
Crankshaft assembly: examination and replacement ... 9	Refitting the engine unit into the frame ... 20
Small end bush: examination and replacement ... 10	Starting and running the rebuilt engine ... 21
Piston and piston rings: examination and renovation ... 11	

Specifications

Engine
Type	Single cylinder, fan-cooled two-stroke
Porting	Crankcase induction
Capacity	49.77 cc (3.03 cu in)
Bore	38.4 mm (1.51 in)
Stroke	43.0 mm (1.69 in)
Compression ratio	9:1
Lubrication	By petrol/oil (petroil) mixture

Piston
Nominal piston/bore clearance	0.105 mm (0.004 in)
wear limit	0.155 mm (0.006 in)

Piston rings
Type	Plain, pegged
Number	Two
End gap	0.1 to 0.25 nominal (0.004 to 0.010 in)
	2.0 mm wear limit (0.078 in)

1 General description

The engine unit fitted to the Vespa Bravo and Ciao mopeds is a horizontal single cylinder two-stroke, employing fan cooling to maintain consistent cooling at all engine speeds.

The crankshaft assembly doubles as a form of rotary valve, the incoming mixture being drawn into a cutout in the flywheel from the crankcase-mounted carburettor.

The piston is fitted with two plain piston rings which are pegged in traditional two stroke fashion to prevent their ends from becoming caught in the port windows and fracturing.

The cylinder head incorporates a decompressor mechanism to facilitate easy starting and stopping of the engine. It takes the form of a small poppet valve, not unlike the inlet and exhaust valves in a four-stroke engine, which is normally held by a spring in the closed position.

The crankcase comprises two castings, one of which is a small cast end plate supporting the right-hand main bearing. The other, left-hand, casting is much larger, and forms part of the cooling ducting. The crankshaft consists of two flywheels between which is carried the connecting rod and big end bearing assembly. The right-hand mainshaft is a short pin serving only to locate and support the assembly in the right-hand main bearing. The left-hand mainshaft protrudes through the left-hand main bearing and carries the generator rotor/fan assembly, and the driving clutch.

The unit is lubricated by oil carried in suspension in the incoming fuel. As the fuel is drawn into the crankcase, the oil content lubricates the various moving parts with which it comes into contact; namely, the big end, small end and main bearings, and the lower part of the cylinder bore. A certain amount of oil is released during the combustion stroke, which acts as an upper cylinder lubricant, the remainder being burnt and expelled along with the exhaust gases.

2 Operations with the engine in the frame

It is not necessary to remove the engine from the frame to accomplish the following operations:
1 Removal and replacement of the cylinder head.
2 Contact breaker checking.
3 Removal and replacement of the carburettor.
4 Removal and replacement of the ignition coil.

3 Operations with the engine removed from the frame

All operations other than those listed in the above Section, entail the removal of the engine from the frame. This is not as daunting as it might seem at first as the engine can be removed in a matter of minutes, with a little practice. It should be noted that the following operations in particular will require the removal of the engine:
1 Removal of the cylinder barrel and piston.
2 Removal of the generator rotor/fan assembly.
3 Removal of the contact breaker assembly.
4 Removal of the crankshaft assembly.

4 Removing the engine from the frame

1 Place the machine securely on its centre stand. If possible, it is recommended that the machine is placed on some sort of raised platform, to make working easier. It is not difficult for one person to lift the machine bodily.
2 Start by removing the outer plastic side covers and the top engine cover. These are retained by Dzus fasteners and a single screw respectively (four screws on the Bravo models). Slacken the exhaust pipe clamp at the manifold, and remove the silencer mounting bolt. The exhaust system can then be lifted clear of the frame. Note that on Ciao models, the silencer is attached to the right-hand side of the machine, and that on Bravo models it is mounted on the left. The method of removal is, however, the same in each case.
3 Slacken the nut at the centre of the centrifugal clutch assembly (single speed models) or speed governor, which is mounted on the left-hand side of the crankshaft. The engine can be prevented from turning by passing a screwdriver through the frame member and between two of the fins on the flywheel. With the nut removed, the clutch and driving belt can be drawn off the mainshaft, followed by the spacer (where fitted).
4 Disconnect the decompressor cable at the cylinder head. A screwdriver can be used to hold the mechanism against the pressure of the return spring while the cable is released. Pull off the two electrical leads from the top of the crankcase, having first made a note of which lead corresponds to which of the two Lucar terminals. Remove the high tension lead from the sparking plug, and thread the lead and cap back through the frame and out of the way of the engine. On Ciao models, remove the top cover mounting plate, which is held by two screws to the right-hand side of the frame section.
5 It is now necessary to release the carburettor and air cleaner. The former need not be removed entirely, but can be tied up out of the way of the engine. The air cleaner is clamped to the carburettor by a single screw, and will probably need to be removed to give sufficient clearance for the carburettor to be detached. On Ciao models, it may prove necessary to wait until the engine bolts are slackened and the engine is pulled forward, to give enough clearance. The carburettor is also clamp mounted to the rear of the crankcase and may be pulled off, once the screw has been slackened.
6 Remove the screws which retain the ignition coil. This is mounted behind the right-hand plastic cover on Bravo models, and beneath the engine unit or under one of the plastic covers on Ciao models. The coil should be tied clear in a similar manner to that of the carburettor. On Bravo models, detach the pressed steel undershield which is fitted between the two engine-supporting frame members.
7 Check round the engine for any remaining connections which might impede engine removal. For example, Bravo models have a cable guide clip which holds the rear brake cable to the fan cowling. The engine is now retained by two bolts which pass through lugs in the front of the crankcase and through both sides of the pressed steel frame members. A third short bolt is screwed into the left-hand side of the unit at the rear.
8 Remove the short rear bolt and remove the nuts from the two front mounting bolts. These can now be pushed through to release the engine unit, which will stay in position between the frame members. Grasp the unit by the cylinder head, and manoeuvre it downwards and out of the frame. It will be found necessary to twist the unit gradually in an anti-clockwise direction, when viewed from the front, in order to disengage the crankshaft from the hole in the frame member through which it passes. The engine can then be lifted clear of the frame and placed to one side, to await further attention.

4.2a Exhaust system is retained by clamp ...

4.2b ... and this bolt through the side member

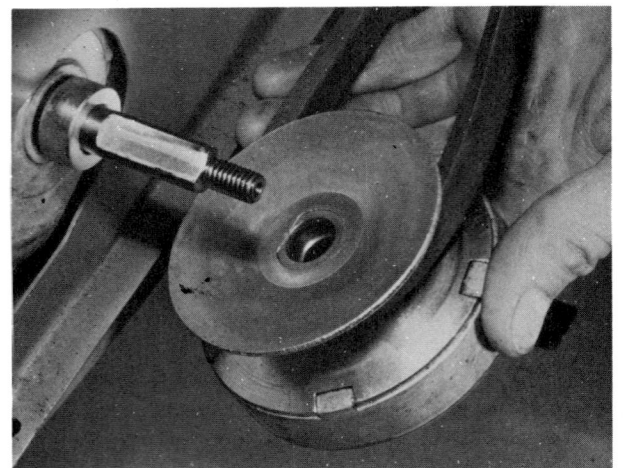

4.3 Remove clutch or speed governor, as applicable

4.6 Detach ignition coil and engine shield

4.7 Remove two front and single rear mounting bolts (arrowed)

4.8a Twist engine during removal to clear mainshaft ...

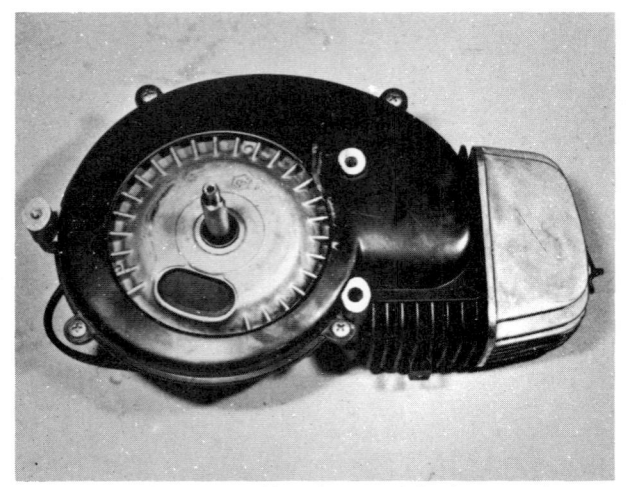

4.8b ... then place on workbench to await further attention

5 Dismantling the engine: general

1 Before commencing work on the engine unit, the external surfaces should be cleaned thoroughly. A motor cycle engine has very little protection from road grit and other foreign matter, which will find its way into the dismantled engine if this simple precaution is not taken. One of the proprietary cleaning compounds, such as 'Gunk' or 'Jizer' can be used to good effect, particularly if the compound is permitted to work into the film of oil and grease before it is washed away. Special care is necessary, when washing down to prevent water from entering the now exposed parts of the engine unit.

2 Never use undue force to remove any stubborn part unless specific mention is made of this requirement. There is invariably good reason why a part is difficult to remove, often because the dismantling operation has been tackled in the wrong sequence.

3 It should be noted at this juncture that two service tools will be required in order to completely dismantle the engine unit. The first is a threaded extractor which must be used to draw the flywheel rotor off its taper. The part number of the tool is T.0035485. The second tool is a crankcase separation tool which presses the crankcase halves apart. Its number is

Chapter 1 Part 1 Engine

T.0035483. These tools may be purchased through an authorised Vespa Service Agent, but in view of the infrequency of their use it is unlikely that it will be worthwhile buying them. If the owner is on good terms with his local Vespa Agent, he may be persuaded to lend or hire the tools over a weekend when he is unlikely to need them himself. Alternatively, the unit can be taken to the Service Agent and he can then remove the rotor and crankcases. (Make prior arrangements for this to be done!).

4 It is our experience that the flywheel/rotor puller is absolutely essential. There is **no** safe alternative method of removal. As a last resort, the use of the crankcase separator may **possibly** be avoided, and this will be discussed in the relevant Section. It is stressed, however, that there is a very real risk of damage if the correct tool is not used.

6 Dismantling the engine: removing the cylinder head, barrel and piston

1 Remove the plastic cylinder cowl, and then slacken the three cylinder head nuts evenly using an 11 mm socket or box spanner. The cylinder head can now be lifted off, noting that no head gasket is fitted.

2 Lift the cylinder barrel carefully up along the studs, taking care to catch the piston as it emerges from the bore. As a precuation against pieces of broken piston ring entering the crankcase mouths, a piece of clean rag should be used to pack around the connecting rod, covering the openings.

3 Remove the circlips from the piston and press out the gudgeon pin so that the piston is released. If the gudgeon pin is a particularly tight fit, the piston should be warmed first, to expand the alloy and release the grip on the steel pin. If it is necessary to tap the gudgeon pin out of position, make sure that the connecting rod is supported to prevent distortion. On no account use excessive force. Discard the old circlips.

4 Note that the piston crown is marked with an arrow which faces down towards the exhaust port. This is a useful guide to ensure that the piston is refitted correctly.

7 Dismantling the engine: removing the flywheel rotor

1 The flywheel rotor fulfills several functions, having fins on the outside to provide forced cooling for the engine, magnetic poles to form the magneto rotor, whilst performing a third function as an engine auxiliary flywheel. The centre boss of the rotor is threaded to accept the extractor tool mentioned in Section 5 of this Chapter. As mentioned earlier, it is essential that this tool is used, as there is no other safe way of removing the rotor. The fact that the rotor runs inside the cast ducting for the cooling system precludes the use of a normal legged puller.

2 Slacken and remove the cheese headed screws which retain the plastic fan ducting to the crankcase. The ducting can then be lifted away. Screw the extractor into the rotor boss, and then gradually tighten the centre bolt until the rotor comes free. It may prove helpful to tap the centre bolt **very lightly** rather than apply excessive pressure. On no account strike the actual flywheel as the fins are easily fractured and there is some risk of demagnetizing the integral rotor.

3 If there is any tendency for the engine to turn during removal, it can be locked by passing a bar through the small end eye and supporting the ends on small wooden blocks against the crankcase mouth. This should not normally be necessary, however, as the taper joint usually separates easily. The rotor can now be lifted away. Note that it is not necessary to disturb the rest of the magneto components, with the exception of the contact breaker cam (which can be slid off), in order to separate the crankcase halves.

6.1a Remove plastic cylinder shroud ...

6.1b ... then slacken cylinder head nuts

6.2 Slide barrel off studs, catching piston as it emerges

6.3 Remove and discard circlips

7.2 Use correct tool to withdraw flywheel rotor

8 Dismantling the engine: separating the crankcase halves

1   Having removed the flywheel rotor as described in the preceding Section, the crankcase securing bolts should be released. Obtain Service tool No. T0035483, and fit it into position on the generator side of the crankcase assembly. Fit and tighten the three knurled screws which clamp the tool to the crankcase.
2   Gradually tighten the T-handle until the crankshaft assembly, complete with the smaller crankcase half, is displaced. The large crankcase half can now be placed to one side. As mentioned previously, the above method is recommended as being the only safe way of separating the crankcases, and it is worth having this job done by a qualified Vespa Service Agent, if necessary. Bear in mind that crankcase separation is normally necessary only in the event of the crankshaft assembly requiring renovation. This is not a job which can easily be undertaken at home as it requires the use of specialised equipment.
3   Wear in the crankshaft assembly can be assessed prior to separation by feeling for play in the big end bearing (see Section 9) and by feeling for roughness in the main bearings as the crankshaft is turned. If wear or play is evident, the best course of action is to entrust the bare crankcase and crankshaft to a Vespa Service Agent. He will then be able to effect separation using the correct tools, and fit a Service Exchange crankshaft. It should be noted that a large proportion of the cost of fitting an exchange crankshaft is in the time involved in removing the engine from the frame, and dismantling up to the point of separation, so a substantial amount will have been saved if the above course of action is taken.
4   It is possible to effect separation without the use of the manufacturer's service tool, by using a conventional large legged puller. This method calls for the use of great care. It is relatively easy to damage the light alloy castings, and this approach is not, therefore, recommended.
5   Assuming that the crankcases have been separated, it will be necessary to remove the smaller, blind, crankcase from the end of the crankshaft assembly. This job can be tackled in two ways, depending on the type of crankshaft fitted. Examine the flywheels. On some crankshaft assemblies, two threaded holes will be found. In this case, obtain two 6 mm bolts of 1 mm pitch thread, and screw these into the holes in the flywheels until they butt against the inside of the casing. Continue screwing the bolts in progressively and evenly to draw the crankshaft assembly free of the casing.
6   On machines fitted with a crankshaft assembly not having the extractor holes, the casing can be removed by heating it until it expands sufficiently to release the bearing. This is best done by placing the assembly in an oven and heating it to 100°C (212°F). The light alloy casing will expand much faster than the steel bearing, and will probably drop away when the assembly is removed. The cover may be tapped lightly, if necessary, to assist removal. The cover should be fitted in the same way, regardless of whether the threaded or plain flywheels are fitted.
Note: Uneven heating of aluminium alloy castings can cause warping due to the differing rates of expansion. For this reason, the use of a blowlamp as a heat source is not recommended, unless the owner has previous experience of this method.

9 Crankshaft assembly: examination and replacement

1   The crankshaft assembly comprises two one-piece full flywheels and crank main pins. These are joined by a press fit central crankpin upon which the big end bearing and connecting rod run. The whole is run in two journal ball bearings, one on each main pin.
2   The main bearings should be washed free of all old oil deposits, as they cannot be properly tested before this is done. If any play is evident or if the bearings do not run freely and smoothly they must be replaced. Warning of main bearing failure is usually given by a characteristic rumble that can be readily heard when the engine is running. Some vibration will also be felt, which will be transmitted via the footrests and frame in general.
3   It is advisable to entrust main bearing renewal to a Vespa Service Agent, who can supply, and if necessary fit, a Service Exchange crankshaft assembly. This applies equally in the case of big end bearing wear, which can be checked as follows:
4   Grasp the connecting rod and feel for any discernible movement. An axial clearance is intentional, but should not exceed 0.65 mm (0.0256 in). Any up and down movement will necessitate renewal of the crankshaft assembly. Because of the method of fixing the central crankpin and of the high accuracy required in re-aligning the flywheels after bearing replacement, this operation is usually beyond the means of the average owner. It is recommended that the crankshaft assembly be returned to a Vespa Agent and a Service Exchange component obtained.

Chapter 1 Part 1 Engine

8.2 Crankcase separation tool in position on unit

8.6 Late type flywheels (as shown) do not have extraction holes

9.3 Obtain exchange crankshaft assembly, if worn

10 Small end bush: examination and replacement

1 Wear in the small end bush is characterised by a light metallic rattle which is particularly noticeable under load. Wear can be checked by inserting the gudgeon pin in the bush and feeling for movement. When new, the bush to gudgeon pin fit should be free from any discernible play, that is, a light sliding fit. The maximum allowable play is 0·02 mm (0·0008 in).

2 Bush renewal is accomplished by using the drawbolt arrangement shown in Fig. 1.2. The new bush will be drawn into position, at the same time displacing the old, worn bush. The new bush will probably have to be reamed out using an expanding reamer to obtain the correct fit; note that the oil hole must be drilled out after fitting. If required, this work can be entrusted to a Vespa Service Agent – it should not prove very expensive.

11 Piston and piston rings: examination and renovation

1 Attention to the piston and rings can be overlooked if a rebore is necessary because new piston and rings will be fitted under these circumstances.

2 If a rebore is not considered necessary, the piston should be examined closely. Reject a piston if it is badly scored or discoloured as the result of the exhaust gases by-passing the rings. Check the gudgeon pin bosses to ensure that they are not enlarged or that the grooves retaining each circlip are not damaged.

3 Remove all carbon from the piston crown and use metal polish to finish off, so that a high polish is obtained. Carbon will adhere much less readily to a polished surface. Examination will show whether the engine has been rebored previously since the amount of overbore is invariably stamped on each piston crown. Three oversizes are available, in 0·20 mm increments.

4 The grooves in which the piston rings locate can become enlarged in use. The clearance between the edge of each piston ring and the groove in which it seats should not exceed 0.05 mm (0.002 in).

5 Remove the piston rings by pushing the ends apart with the thumbs whilst gently easing each ring from its groove. Great care is necessary throughout this operation because the rings are brittle and will break easily if overstressed. If the rings are gummed in their grooves, three strips of tin can be used, to ease them free, as shown in the accompanying illustration.

6 Piston ring wear can be checked by inserting the rings one at a time in the cylinder bore from the top and pushing them down about 1½ inches with the base of the piston so that they rest squarely in the bore. Make sure that the end gap is away from any of the ports. If the end gap is within the range 0·1 – 0·125 mm (0·004 – 0·005 in) the ring is fit for further service.

7 Examine the working surface of each piston ring. If discoloured areas are evident, the ring should be renewed because these areas indicate the blow-by of gas. Check that there is not a build-up of carbon on the back of the ring or in the piston ring groove, which may cause an increase in the radial pressure. A portion of broken ring affords the best means of cleaning out the piston ring grooves.

8 Check that the piston ring pegs are firmly embedded in each piston ring groove. It is imperative that these retainers should not work loose, otherwise the rings will be free to rotate and there is danger of the ends being trapped in the ports.

9 It cannot be over-emphasised that the condition of the piston and piston rings is of prime importance because they control the opening and closing of the ports by providing an effective moving seal. A two-stroke engine has only three working parts, of which the piston is one. It follows that the efficiency of the engine is very dependent on the condition of piston and the parts with which it is closely associated.

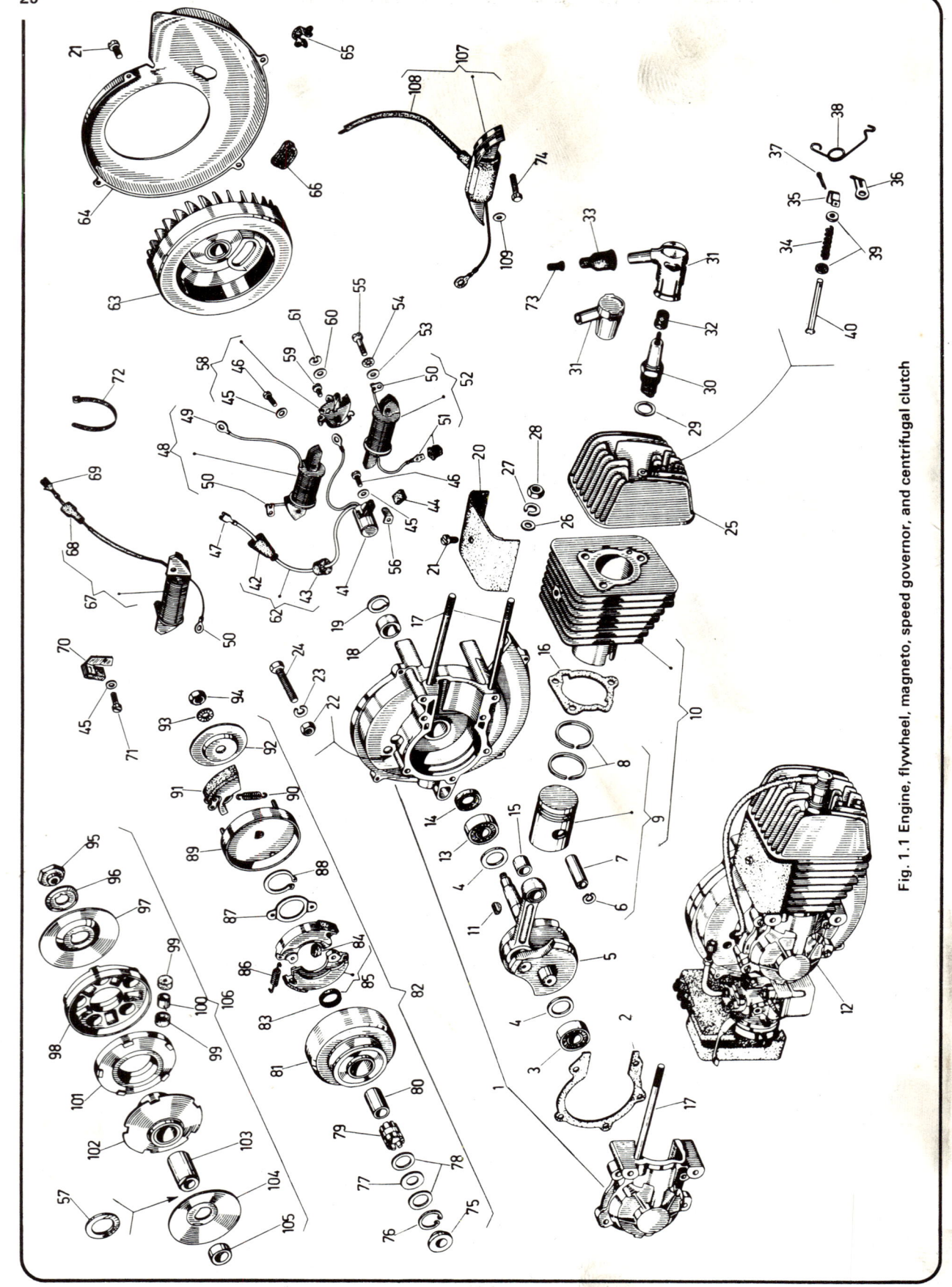

Fig. 1.1 Engine, flywheel, magneto, speed governor, and centrifugal clutch

1 Crankcase assembly
2 Crankcase gasket
3 Right-hand main bearing
4 Thrust washer – 2 off
5 Crankshaft assembly
6 Circlip – 2 off
7 Gudgeon pin
8 Piston ring set
9 Piston assembly
10 Cylinder and piston assembly
11 Woodruff key
12 Complete engine unit
13 Left-hand main bearing
14 Crankshaft seal
15 Small end bush
16 Cylinder base gasket
17 Stud – 3 off
18 Contact breaker cam
19 Thrust washer
20 Cylinder cowl
21 Screw – 5 off
22 Nut – 5 off
23 Spring washer – 5 off
24 Bolt – 5 off
25 Cylinder head
26 Plain washer
27 Spring washer – 4 off
28 Nut – 3 off
29 Sparking plug sealing washer
30 Sparking plug
31 Sparking plug cap
32 Seal
33 Gaiter
34 Spring
35 Clip
36 Anchor plate
37 Split pin
38 Hair spring
39 Spring seat – 2 off
40 Decompressor valve
41 Condenser
42 Dust seal
43 Grommet
44 Lubricating wick
45 Plain washer – 2 off
46 Screw
47 Lucar terminal
48 Ignition low tension coil
49 Terminal – 3 off
50 Terminal – 2 off
51 Insulating block
52 Lighting coil
53 Plain washer – 2 off
54 Shakeproof washer – 4 off
55 Screw – 4 off
56 Bracket
57 Thrust washer
58 Contact breaker assembly
59 Screw
60 Plain washer – 3 off
61 Spring clip
62 Low tension lead
63 Flywheel/rotor
64 Main cowling
65 Cable guide
66 Rubber plug
67 Alternative lighting coil
68 Terminal cover
69 Lucar terminal
70 Lubricating wick
71 Retaining screw
72 Cable tie
73 Sleeve
74 Mounting bolt – 2 off
75 Spacer
76 Circlip
77 Shouldered washer
78 Washer – 2 off
79 Caged roller bearing
80 Inner bush
81 Pulley/clutch drum
82 Centrifugal clutch assembly
83 Seal
84 Rubber stop – 2 off
85 Starting shoe assembly
86 Return spring – 2 off
87 Retaining plate
88 Circlip
89 Back plate/starting shoe drum
90 Return spring – 3 off
91 Main clutch shoe – 3 off
92 End plate
93 Shakeproof washer
94 Nut
95 Special nut
96 Tab washer
97 Cover plate
98 Speed governor body
99 Nylon roller sleeve – 10 off
100 Roller – 5 off
101 Roller housing half
102 Pulley half
103 Bush
104 Pulley half
105 Spacer
106 Centrifugal speed governor assembly
107 Alternative ignition low tension coil
108 HT cable
109 Plain washer – 4 off

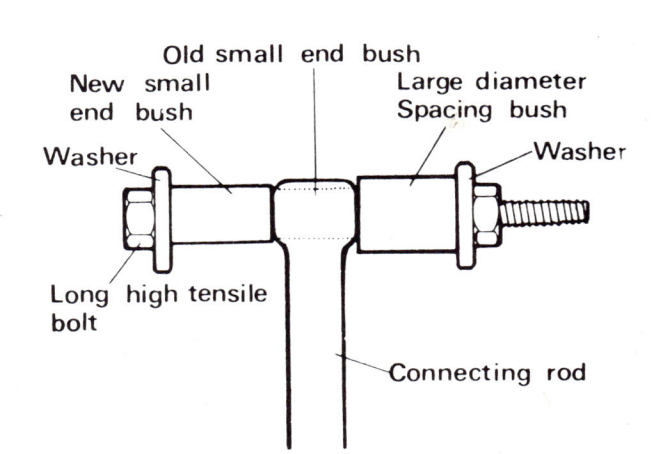

Fig. 1.2 Using drawbolt method to renew small-end bush

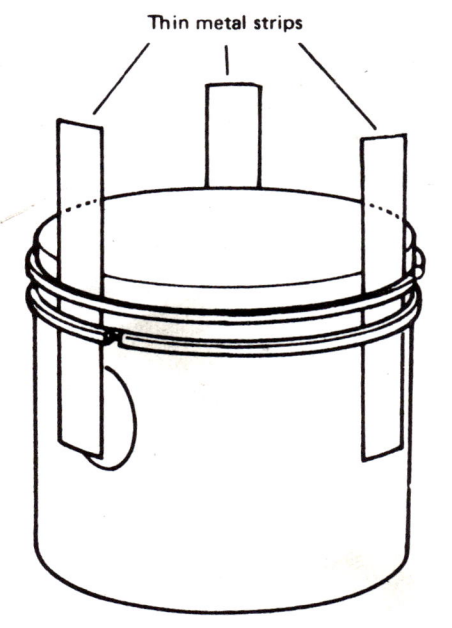

Fig. 1.3 Freeing gummed rings

12 Cylinder barrel: examination and renovation

1 The cylinder barrel should be carefully cleaned using a wire brush and petrol to remove any accumulation of grime around the cooling fins. After drying the bore with a clean rag, examine the surface for signs of wear or scoring. If scoring or scratches are in evidence in the bore, the cylinder barrel will need to be rebored and a new piston fitted.

2 A small ridge may be in evidence near to the top of the bore. This marks the extent of travel of the top piston ring, and will probably be more pronounced at one point (the thrust face) than at any other. If this is barely perceptible, and the piston and rings are in good condition, it will probably be safe to use the existing bore. If in any doubt, and in any case if the ridge is marked, the barrel should be taken to a Vespa Service Agent for checking, together with its piston.

3 In the event that a rebore is necessary, it should be noted that Vespa Service Agents are able to supply rebored barrels with new, matched pistons as a Service Exchange item. This is an economical and quick way of reconditioning the engine. Pistons and cylinders supplied by the factory have letters stamped on them to denote size. Any barrel and piston combination must share a common sizing number.

4 For those owners possessing facilities for cylinder bore and piston measurements, there follows a list of these dimensions for each rebore size:

Oversize	Piston diameter	Bore size
Standard	38.290-38.320 mm	38.395-38.425 mm
1st oversize	38.495-38.515 mm	38.600-38.620 mm
2nd oversize	38.695-38.715 mm	38.800-38.820 mm
3rd oversize	38.895-38.915 mm	39.000-39.020 mm

Nominal piston to cylinder bore clearance is: 0.105 mm (0.0041 in). Maximum wear limit is: 0.155 mm (0.0061 in).

5 Clean all carbon deposits from the exhaust ports using a blunt ended scraper. It is important that all the ports should have a clean, smooth appearance because this will have the dual benefit of improving gas flow and making it less easy for carbon to adhere in the future. Finish off with metal polish, to heighten the polishing effect.

6 Do not under any circumstances enlarge or alter the shape of the ports under the mistaken belief that improved performance will result. The size and position of the ports predetermines the characteristics of the engine and unwarranted tampering can produce very adverse effects.

13 Cylinder head: examination and renovation

1 It is unlikely that the cylinder head will require any special attention apart from removing the carbon deposit from the combustion chamber. Finish off with metal polish; the polished surface will help improve gas flow and reduce the tendency of future carbon deposits to adhere so readily.

2 Check that the cooling fins are clean and unobstructed, so that they receive the full air flow.

3 Check the condition of the thread within the sparking plug hole. The thread is easily damaged if the sparking plug is overtightened. If necessary, a damaged thread can be reclaimed by fitting a Helicoil thread insert. Most Vespa Agents have facilities for this type of repair, which is not expensive.

4 If there has been evidence of oil seepage from the cylinder head joint when the machine was in use, check whether the cylinder head is distorted by laying it on a sheet of plate glass. Severe distortion will necessitate renewal of the cylinder head, but if distortion is only slight, the head can be reclaimed by wrapping a sheet of emery cloth around the glass and using it as the surface on which to run down the head with a rotary motion, until it is once again flat. The usual cause of distortion is failure to tighten down the cylinder head nuts evenly, in a diagonal sequence.

5 A decompressor valve assembly is fitted to the cylinder head. When opened, it allows the cylinder compression to pass straight into the exhaust system. The purpose of this is to allow the pedals to be turned quickly to facilitate starting, and it also acts as a means of stopping the engine. The valve will rarely require attention, but should there be evidence of leakage, proceed as follows:

6 Remove the split pin which retains the saddle to the top of the valve stem. The large hairspring can be lifted away together with the saddle, as can the valve spring and its upper and lower seats. The valve can then be pushed inwards and removed from the combustion chamber.

7 Examine the valve and valve seat faces for signs of blow-by or leakage. If the sealing surfaces appear inadequate, they should be ground in the same way as those in four-stroke engines. Smear a little fine grinding paste on the valve seat, and place the valve in position in the head. The two surfaces should be lapped together with a to and fro motion, using a screwdriver on the slot in the head. The valve should be lifted and turned every two or three oscillations, to prevent grooves being formed by the abrasive particles. When finished, the valve and valve seat faces should have a uniform dull grey surface with no signs of scores or other marks. Ensure that all traces of the grinding paste are washed out prior to reassembly.

8 Reassembly is a direct reversal of the removal sequence. Note that some difficulty may be encountered in keeping the valve in place while the springs and split pin are refitted. It was found that a woodscrew, screwed part way into the workbench, makes a suitable support for the valve head.

14 Engine reassembly: general

1 Before reassembly of the engine unit is commenced, the various component parts should be cleaned thoroughly and placed on a sheet of clean paper, close to the working area.

2 Make sure all traces of old gaskets have been removed and that the mating surfaces are clean and undamaged. One of the best ways to remove old gasket cement is to apply a rag soaked in methylated sprit. This acts as a solvent and will ensure that the cement is removed without resort to scraping and the consequent risk of damage.

3 Gather together all of the necessary tools and have available an oil can filled with clean engine oil. Make sure all the new gaskets and oil seals are to hand, also all replacement parts required. Nothing is more frustrating than having to stop in the middle of a reassembly sequence because a vital gasket or replacement has been overlooked.

15 Engine reassembly: joining the crankcase halves

1 The crankshaft assembly should first be installed in the larger of the two crankcase halves. The crankcase bearing boss should be warmed to expand the alloy which in turn will allow the bearing to fit easily. This is best done by playing a blowlamp flame around the boss from the inside. There should be no need to remove the flywheel generator components during this process, as the heat build up should not be that great. The flame should be kept moving to ensure that the boss is heated evenly. If this precaution is not observed, there is a risk of the casing becoming warped. The crankshaft assembly can now be placed into position, and the casing allowed to cool.

2 Heat the smaller casing half in an oven to a temperature of 100°C (212°F) to ensure that it will slip easily over the bearing. Fit a new gasket on the mating surface of the larger crankcase half, then fit the heated smaller half into position, ensuring that it is tapped fully home. Refit the crankcase securing bolts and nuts, and tighten them progressively. Ensure that the crankshaft turns easily when reassembly is complete. Any tight spots are an indication that something has been assembled incorrectly, and this must be eradicated before any further reassembly takes place.

Chapter 1 Part 1 Engine

13.6a Withdraw split pin to release saddle and hairspring ...

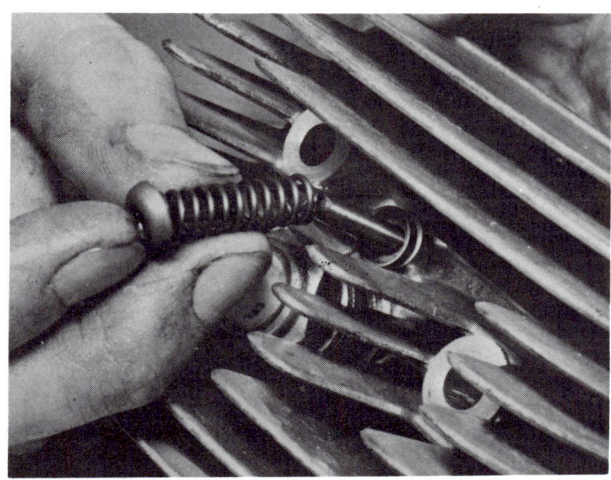

13.6b ... then remove valve spring and seals

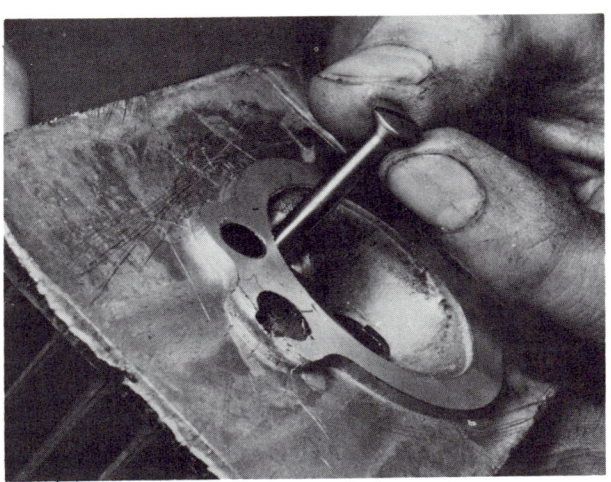

13.6c Valve can be removed from cylinder head

15.1 Fit new crankshaft seal, then crankshaft assembly

15.2 Warm smaller crankcase half before fitting

16 Engine reassembly: replacing the piston

1 Before the piston is refitted, cover the crankcase opening with rag to obviate any risk of a displaced circlip entering the crankcase. The piston has an arrow stamped on the crown which must point down towards the exhaust port when fitted.

2 If the gudgeon pin is a tight fit in the piston boss, the piston can be warmed with warm water to effect the necessary temporary expansion. Oil the gudgeon pin and piston boss before the gudgeon pin is inserted, then fit the circlips, making sure that they are engaged fully with their retaining grooves. A good fit is essential, since a displaced circlip will cause extensive engine damage. Always fit new circlips, NEVER re-use the old ones.

3 Check that the piston rings are fitted correctly, with their ends either side of the ring pegs.

If this precaution is not observed, the rings will be broken during assembly.

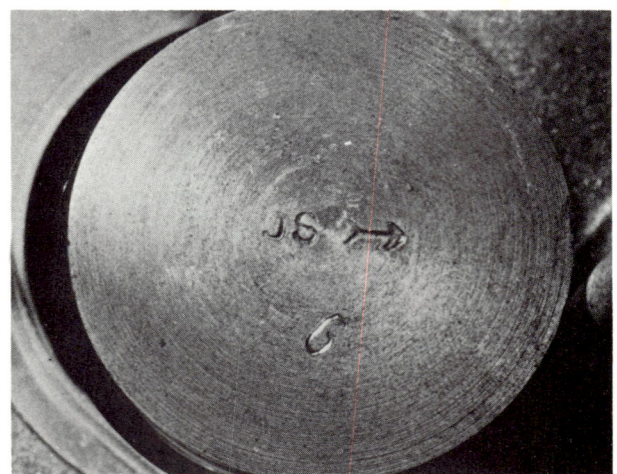

16.1 Arrow **must** face downwards

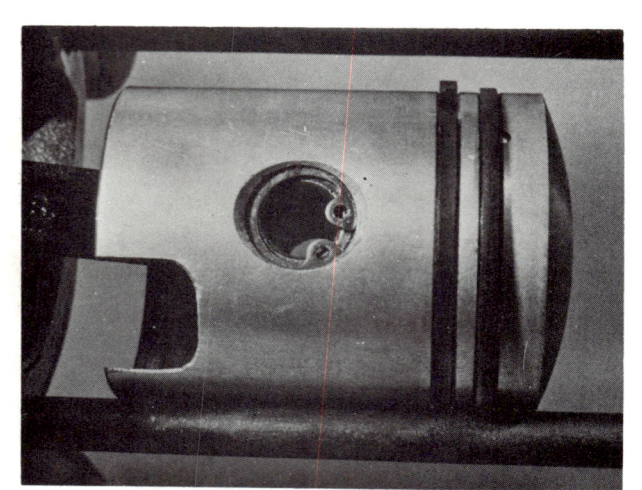

16.2 Insert gudgeon pin ...

16.2b ... and fit two new circlips

16.3 Rings must be positioned as shown

17 Engine reassembly: replacing the cylinder barrel

1 Place a new cylinder base gasket over the retaining studs and lubricate the cylinder bore with clean engine oil. Arrange the piston so that it is at top dead centre (TDC) and lower the cylinder barrel over the retaining studs until it reaches the piston crown. The rings can now be squeezed one at a time until the cylinder barrel will slide over them, checking to ensure that the ends are still each side of the ring peg. Great care is necessary during this operation, since the rings are brittle and very easily broken.

2 When the rings have engaged fully with the cylinder bore withdraw the rag packing from the crankcase mouth and slide the cylinder barrel down the retaining studs, so that it seats on the new base gasket (no gasket cement).

18 Engine reassembly: refitting the cylinder head

1 Slide the cylinder head over the retaining studs. The cylinder head cannot be fitted incorrectly as there are only three studs. Ensure that the decompressor hair spring end, and the decompressor cable anchor plate are fitted to their respective studs (see photograph). Fit the retaining nuts and washers, and tighten the nuts evenly and progressively, using an 11 mm socket or box spanner. Make sure that the sparking plug is fitted, to preclude any risk of foreign matter entering the cylinder.

19 Engine reassembly: refitting the flywheel rotor and fan ducting

1 Place one or two drops of light machine oil on the contact breaker cam lubricator wick, and slide the cam into position over the mainshaft. Ensure that the cam locates correctly over its Woodruff key. Place the special wave washer into position, noting that it too has a cutout which engages with the Woodruff key.

2 Lower the flywheel rotor into position over the mainshaft, making sure that the rotor boss engages with the Woodruff key.

3 Fit the small metal ducting plate into position and tighten the single retaining screw. Next, fit the large plastic fan duct onto the crankcase, and fit and tighten the securing screws.

Chapter 1 Part 1 Engine

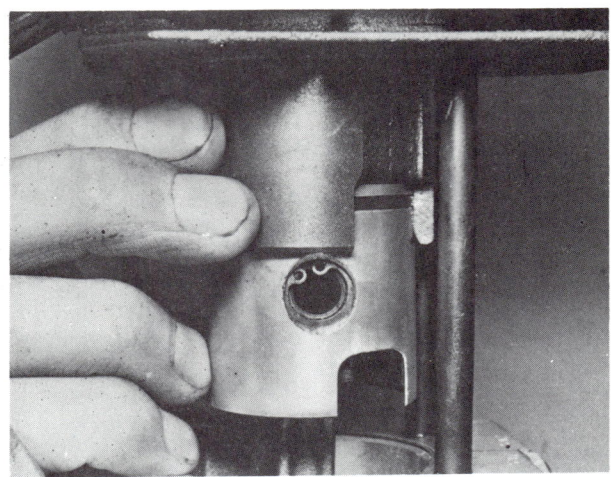

17.1 Feed rings into cylinder bore **carefully**

18.1a Lower cylinder head onto studs ...

18.1b ... Note that cable anchor and spring are held by nuts

19.1 Do not omit this wave washer – Note cutout

20 Refitting the engine unit into the frame

1 Engine refitting is a straightforward reversal of the removal sequence. Place the unit into position between the frame members, passing the crankshaft end through its aperture in the left-hand frame member. Fit the two long front mounting bolts and assemble the nuts and washers loosely. Fit the short rear mounting bolt but do not tighten at this stage. Note that where fitted, the small engine positioning lever should be fitted at this stage.

2 Push the engine fully forward, and refit the carburettor to its stub and tighten the retaining clamp. The air cleaner should be fitted to the carburettor in a similar manner. Reconnect the fuel pipe and fit the carburettor top, if these have been removed. Check, and if necessary adjust, the free play in the throttle cable. There should be about 3 mm ($\frac{1}{8}$ in) at the adjuster.

3 Reconnect the two Lucar connectors to their terminals on the crankcase. Refit the ignition coil, together with the engine undershield (where fitted). Hook the nipple of the decompressor cable to its anchor plate, and position the outer cable in its fork in the hairspring.

4 Pull the engine unit fully rearwards, fit the stepped distance washer with the reduced diameter outwards, and slide the automatic clutch/drive pulley unit onto the end of the mainshaft, with the drive belt already in position. Adjust the belt tension by moving the unit forwards until the correct setting is obtained (see Section 3 of the six monthly/2500 mile service interval in the Routine Maintenance Section of this Manual). The two long, and single short, engine mounting bolts can now be tightened.

5 Fit the exhaust pipe over the exhaust stub and tighten the clamp. Fit and tighten the silencer mounting bolt. On Ciao models, position the top cover mounting plate, and fit the two retaining bolts. Replace the top cover on all models, noting that it is retained by a single, central screw, on Ciao models, and by four smaller screws on Bravo models.

21 Starting and running the rebuilt engine

1 When the initial start-up is made, run the engine slowly for the first few minutes, especially if the engine has been rebored

or a new crankshaft fitted. Check that all the controls function correctly and that there are no oil leaks, before taking the machine on the road. The exhaust will emit a high proportion of white smoke during the first few miles, as the excess oil used whilst the engine was reassembled is burnt away. The volume of smoke should gradually diminish until only the customary light blue haze is observed during normal running. It is wise to carry a spare sparking plug during the first run, since the existing plugs may oil up due to the temporary excess of oil.

2   Remember that a good seal between the piston and the cylinder barrel is essential for the correct function of the engine. A rebored two-stroke engine will require more carefully running-in, over a longer period, than its four-stroke counterpart. There is a far greater risk of engine seizure during the first hundred miles if the engine is permitted to work hard.

3   Do not tamper with the exhaust system or use a holed or damaged silencer. Unwarranted changes in the exhaust system will have a very marked effect on engine performance, invariably for the worse. The same advice applies to dispensing with the air cleaner.

4   Do not on any account add extra oil to the petroil under the mistaken belief that a little extra oil will improve the engine lubrication. Apart from creating excess smoke, the addition of oil will make the mixture much weaker, with the consequent risk of overheating and engine seizure.

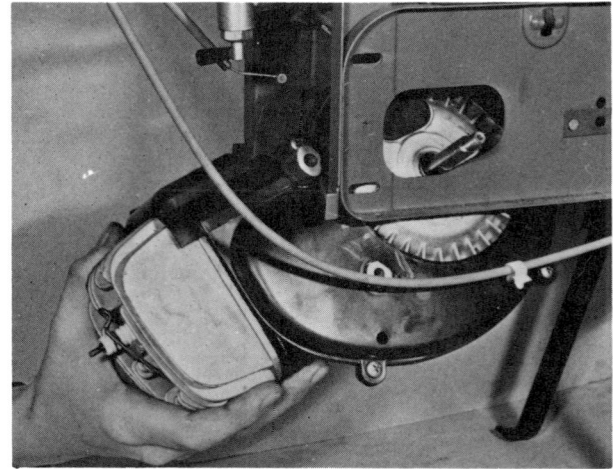

20.1 Feed engine into position as shown above

Chapter 1 Part 2 Transmission

General description ... 22	Automatic speed governor: removal, examination and renovation ... 28
Clutch unit removal, examination and renovation: single speed models ... 23	Automatic clutch unit: removal, examination and renovation: variable ratio models ... 29
Rear pulley examination: single speed models ... 24	Chainwheel, pedals and pedal shaft: all models ... 30
Rear hub unit: removal, examination and renovation: single speed models ... 25	Drive belt adjustment ... 31
Rear hub unit: removal, examination and renovation: variable ratio models ... 26	Fault diagnosis: engine ... 32
Pedal chain and freewheel unit: all models ... 27	Fault diagnosis: transmission ... 33

Specifications

	Single-speed models	Variable ratio models
Front pulley	Fixed size, in unit with centrifugal clutch	Variable size, in unit with centrifugal speed governor
Rear pulley	Fixed size, direct drive to hub unit	Variable size, spring loaded in unit with centrifugal clutch
Rear hub unit	Single speed, direct drive to wheel	Single speed, drive to rear wheel via idler gears
Method of engine disconnection	Push button on rear hub	Push button on rear hub
Drive coupling	Vee belt	Vee belt

22 General description

Two systems of transmission have been employed on the Ciao and Bravo models, giving an alternative of a basic single speed arrangement or a variable ratio system designed to give better acceleration and hillclimbing ability.

On single speed models, drive from the crankshaft passes to a centrifugal clutch unit. The clutch contains three spring loaded shoes mounted on a backplate driven by the crankshaft and contained in a drum. At tickover, the spring tension is suffiient to hold the shoes in, against their stops, and so no drive is transmitted to the drum and pulley. As the speed of the engine rises, the shoes are thrown outwards by centrifugal force until they begin to rub on the drum. From this point, where the shoes are merely rubbing on the drum surface, increasing engine speed leads to greater pressure from the shoes on the drum until the drive between the crankshaft and pulley is solid.

A second, smaller, clutch is incorporated in the unit, having just two lightly sprung shoes. This operates in the same way as the main unit, but transmits movement from the pedals, via the rear wheel and hub, to the engine for starting purposes. As soon as the engine starts and pedal movement ceases, the drive to the engine is disconnected.

Drive from the front pulley is transmitted by way of a vee-belt to the fixed rear pulley. The rear pulley is mounted on the input shaft of the hub, from which drive is transmitted by way of a small integral pinion to a large pinion and thence to the rear wheel spindle.

On variable speed models, the crankshaft drives a centrifugal speed governor which incorporates a variable diameter pulley. At rest, the two pulley halves are at their furthest point apart, making the effective diameter of the pulley as small as possible. The rear pulley is also of variable diameter and is spring loaded to maintain belt tension. At rest, the two halves are pushed together to give the largest possible pulley diameter. This combination of pulley sizes gives a low overall gearing, providing improved acceleration from rest when compared with the fixed ratio models.

As engine speed rises, steel rollers inside nylon sleeves are forced outwards from the centre of the speed governor. This causes the outer half of the front pulley to move inwards, changing its effective size. As this happens, the rear pulley is forced apart against spring tension, thereby reducing its effective diameter. The result is that the overall gearing is raised, giving a higher road speed at a lower engine speed.

The drive is transmitted to a centrifugal clutch which is very similar to that of the single speed models. In this instance it is carried on the input shaft of the rear hub unit. The hub unit is also similar, but incorporates a pair of idler gears to give clearance for the clutch unit.

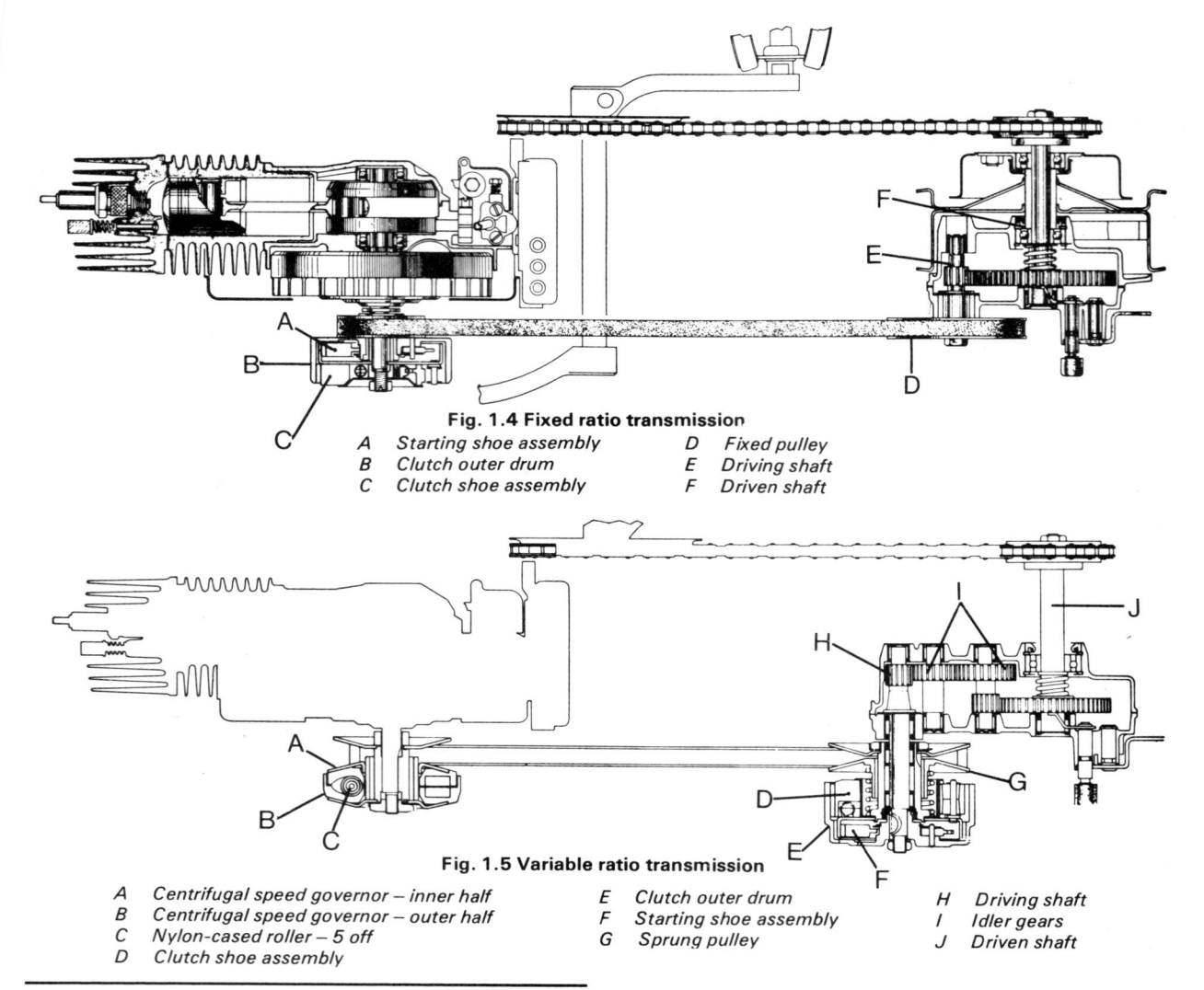

Fig. 1.4 Fixed ratio transmission

A Starting shoe assembly
B Clutch outer drum
C Clutch shoe assembly
D Fixed pulley
E Driving shaft
F Driven shaft

Fig. 1.5 Variable ratio transmission

A Centrifugal speed governor – inner half
B Centrifugal speed governor – outer half
C Nylon-cased roller – 5 off
D Clutch shoe assembly
E Clutch outer drum
F Starting shoe assembly
G Sprung pulley
H Driving shaft
I Idler gears
J Driven shaft

23 Clutch unit removal, examination and renovation: single speed models

1 The clutch unit is located beneath the left-hand outer cover and is carried on the end of the crankshaft. Release the Dzus fasteners which secure the outer cover, and lift it away. Slacken the clutch securing nut. It may prove necessary to lock the engine, and this can be accomplished by inserting a screwdriver blade in the flywheel rotor fins via the cutaway in the frame member. Take care not to damage the rotor fins.

2 Lift the clutch assembly off the end of the crankshaft. The pulley will come away at the same time, together with the drive belt. Disengage the belt and place the clutch unit to one side, to await further dismantling.

3 The clutch shoe assembly and backplate can be lifted out of the drum as a unit. The three shoes are mounted on pivot pins, and can be lifted off as a group. The two inner (starting) shoes are retained by a circlip and locating plate, and can be withdrawn after these have been removed.

4 Examine the drum face for signs of corrosion or contamination. The drum should be carefully washed out in petrol, and if necessary, the drum surfaces polished with fine emery cloth, to obtain a good finish. The clutch shoe backplate is dished inwards to form a smaller drum upon which the starting shoes operate. This too should be cleaned in petrol and the rubbing surface cleaned up, as described above.

5 Examine the shoes of both the main clutch and starting clutch. If badly worn or contaminated with oil or grease, they must be renewed. Normally, however, it is sufficient to clean the shoes and to break the glaze which will have formed on the lining surfaces, using emery cloth. Make sure that the pivot pins are clean and lightly greased, and that the springs are in good order, prior to reassembly.

6 Reassembly is a reversal of the dismantling sequence, bearing in mind the following points; Ensure that the starter shoe retaining plate is positioned correctly over the ends of the pivot pins, and that the circlip is seated securely in its groove. Assemble the main clutch shoes on the backplate (grease the pivot pins lightly) and place the assembly in the main drum. Fit the belt to the drive pulley, and slide the assembly onto the mainshaft, ensuring that the flat in the backplate centre lines up with that on the shaft. Fit the front retainer plate, then fit the shakeproof washer and tighten the securing nut.

24 Rear pulley examination: single speed models

1 The rear pulley on all single speed models is a straightforward fixed size type, and will require no maintenance. Should wear or damage be evident, it will be necessary to remove the pulley for renewal. It is retained by a single, central, retaining nut and washer.

23.2 Remove centre nut to release clutch assembly

23.3a Main shoe assembly can be lifted out of drum

23.3b Remove the circlip and ...

23.3c ... retaining plate to free starting shoes

23.4a Examine and clean housing, and grease pivots

23.4b Backplate forms drum for starting shoes

23.5a Check condition of main clutch shoes ...

23.5b ... and starting clutch shoes

24.1 Plain pulley is used on single-speed models

25 Rear hub unit: removal, examination and renovation: single speed models

1   The rear hub unit is a self-contained assembly housed within the rear wheel hub. It is necessary, therefore, to remove the rear wheel in order to gain access to the unit.

2   Place the machine securely on its centre stand and detach both plastic side covers. Loosen the pedal chain jockey wheel by slackening the nut on the inside of the right-hand frame member. Move the jockey wheel away from the chain to allow it to be disengaged from the freewheel unit on the back wheel. Note that the chain is fitted with a joining link, but it is not normally necessary to remove it in order to separate the chain from the freewheel.

3   Slacken the two front and single rear engine mounting bolts and move the engine rearwards, to allow the drive belt to be lifted off the rear pulley. Detach the rear brake cable at the wheel end. The wheel and hub unit are held to the frame members by two bolts at each side. Remove the bolts and pull the wheel backwards. The hub unit must be turned clockwise (viewed from the left-hand side) as the wheel is lowered so that it clears the projecting end of the left-hand frame member.

4   Using a large thin-jawed spanner, remove the freewheel unit from the right-hand side of the wheel, followed by the mounting plate. The hub unit can then be drawn out of the recess in the left-hand side of the wheel for further dismantling.

It will be noticed that part of the recess forms the rear brake drum, and that the single rear brake shoe is carried on the back of the hub unit.

5   Lock the rear pulley by passing a small rod through one of the holes provided, and remove the securing nut. The pulley can then be pulled off its spindle. It may be considered worthwhile removing the rear brake shoe to avoid any risk of oil contamination. Alternatively, cover the friction material with masking tape.

6   Slacken and remove the four hexagon headed bolts which retain the outer cover to the hub unit body. Lift the cover away carefully, allowing the components to remain on the outer plate, if possible.

7   Examine the pinion teeth for any signs of damage or excessive wear. It is unusual for these components to wear to any marked degree, but if this should be the case, the component(s) concerned must be renewed. There is no satisfactory way of reclaiming damaged parts. Providing the unit has been kept properly lubricated, there should be little evidence of wear in the needle roller bearings in the hub castings. If they appear worn, it is recommended that the hub castings are taken to a Vespa Service Agent who will be able to check whether renewal is necessary, and fit the new bearings, if required. The shafts which run in the bearings may also require renewal, and should also be taken.

8   It will be noticed that the large driven pinion is able to slide on its shaft, and is under light sideways pressure from a spring. A radial slot in the pinion boss engages with a corresponding pin in the shaft, effectively locking the two together. If, for any reason, this pin breaks in use, drive will be lost, although the possibility of this occurring is remote. A fork arrangement is provided by which the pinion can be pushed out of mesh with the driving pinion. In this way, drive to and from the engine is interrupted, allowing the machine to be ridden like a bicycle, if required. Check that the spring which returns the pinion is intact. If broken or weak, the driven pinion may tend to wander out of mesh when the machine is in use.

9   When reassembling the unit, always fit a new housing gasket to ensure oil tightness. Assemble the components on the cover, and then fit this into the main housing. Fit and tighten the securing bolts. Before fitting the unit into the wheel, make sure that the masking tape fitted to the brake shoe has been removed. The wheel should be refitted by reversing the removal sequence described at the beginning of this Section. Refill the hub unit to the level of the filler plug, which will be found just below the button for disengaging the drive. It is worth purchasing the oil in a plastic squeeze pack, as these are ideal for introducing the oil into the filler hole. The hub unit contains SAE EP90 gear oil. This extreme pressure oil should always be used – engine oil will not suffice.

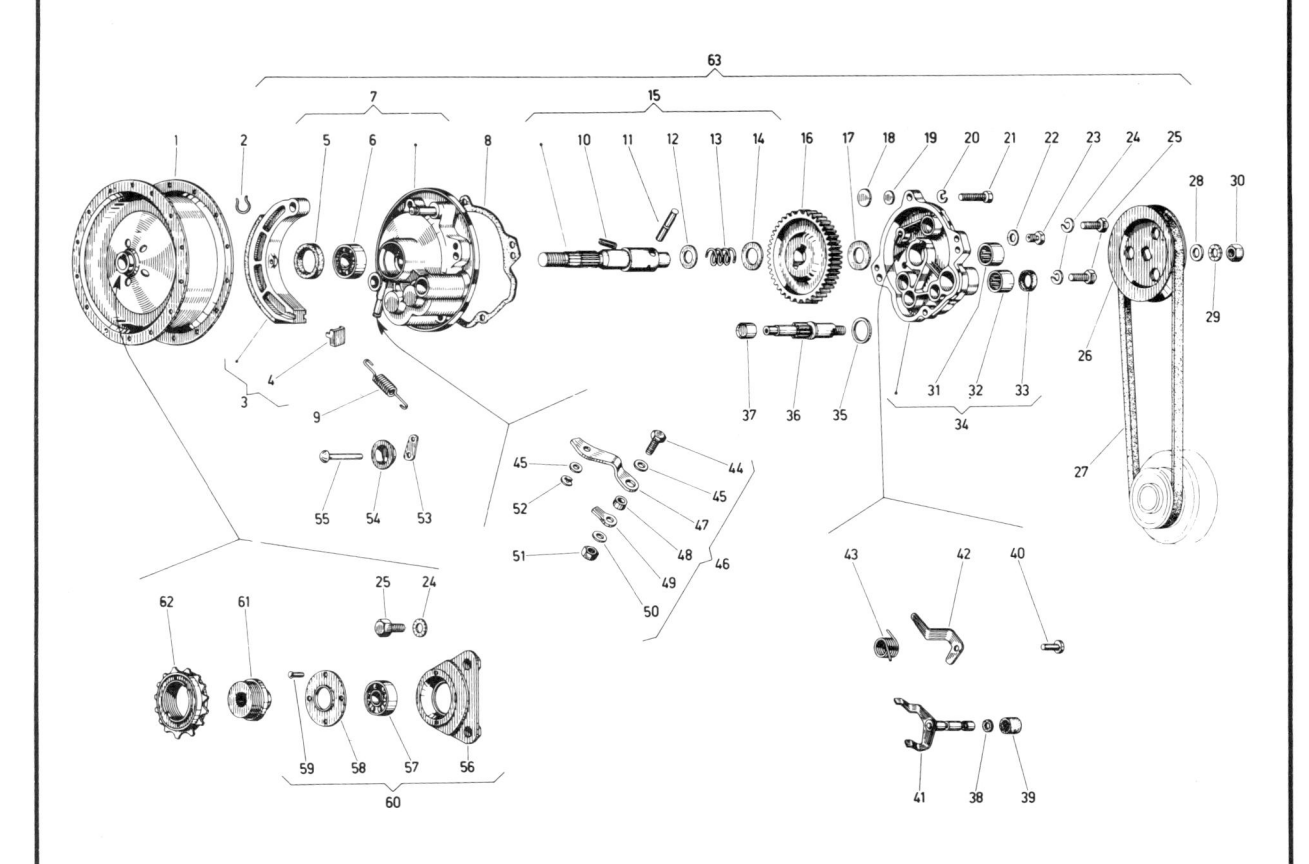

Fig. 1.6 Rear hub unit – fixed ratio models

1 Rear wheel hub
2 Spring clip
3 Brake shoe assembly
4 Steel shoe
5 Oil seal
6 Roller bearing
7 Hub casing assembly
8 Gasket
9 Brake shoe return spring
10 Roll pin
11 Engagement pin
12 Thrust washer
13 Return spring
14 Washer
15 Driven shaft assembly
16 Driven gear
17 Washer
18 Blanking plug
19 Blanking plug
20 Spring washer – 4 off
21 Bolt – 4 off
22 Fibre washer
23 Filler/level plug
24 Spring washer – 4 off
25 Bolt – 4 off
26 Rear pulley
27 Drive belt
28 Washer
29 Shakeproof washer
30 Nut
31 Cup bearing
32 Cup bearing
33 Oil seal
34 Outer cover assembly
35 Washer
36 Driving shaft
37 Cup bearing
38 Fibre washer
39 Plastic cover
40 Pivot pin
41 Disengagement claw
42 Locking lever
43 Lever spring
44 Bolt
45 Plain washer – 2 off
46 Brake actuating lever assembly
47 Brake actuating lever
48 Spacer
49 Cable anchor plate
50 Plain washer
51 Nut
52 Circlip
53 Spring plate
54 Cup
55 Rivet
56 Bearing housing
57 Ball bearing
58 Retaining plate
59 Rivet – 2 off
60 Bearing housing assembly
61 Adaptor
62 Freewheel assembly

25.3 Wheel is retained by two bolts at each side

25.4a Large spanner is required to release freewheel unit

25.4b Hub unit can then be withdrawn from wheel

25.9 Top up hub with SAE EP90 gear oil

26.1a Components should remain in hub cover

26 Rear hub unit: removal, examination and renovation: variable ratio models

1 The variable ratio models utilise a similar type of hub unit to that employed on the fixed ratio models, differing only in the respect it employs two additional idler gears. Remove the wheel and extract the hub unit as described in Section 25, paragraphs 1 to 5. Remove the outer cover of the unit, leaving the hub components in place on it. The two idler gears will be seen to transmit movement from the driving shaft to the driven shaft, and run in bearings similar to the latter. Examination and renovation is to all practical purposes the same as that described in Section 25. Refer to the accompanying photographs for further details of dismantling and reassembly.

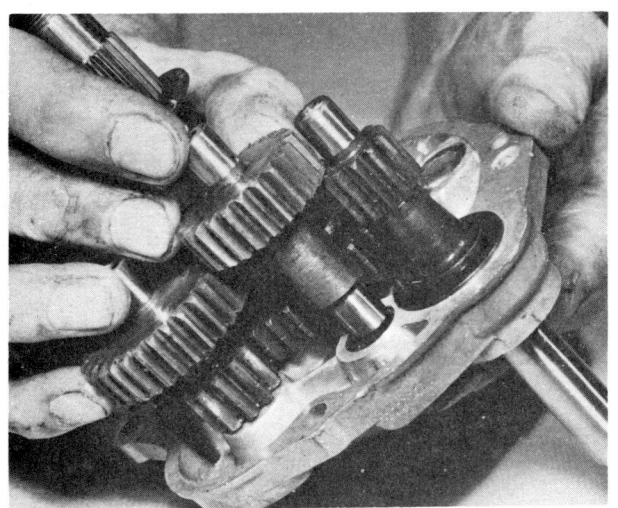

26.1b Remove pinions for examination

26.1c Check the condition of the various bearings ...

26.1d ... and oil seals in the hub castings

26.1e This mechanism operates ...

26.1f ... a fork arrangement which disengages ...

26.1g ... the sliding driven shaft pinion

34

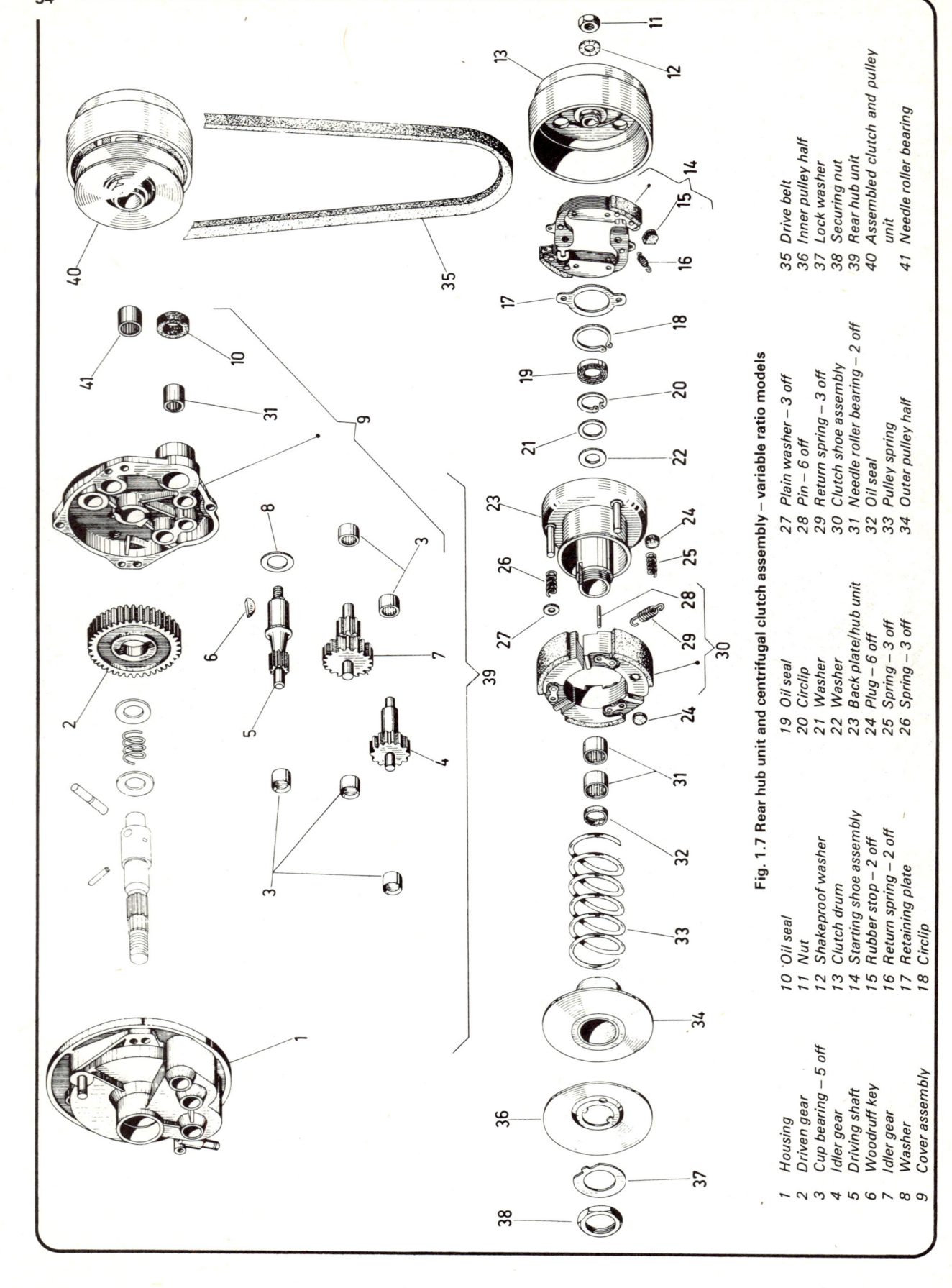

Fig. 1.7 Rear hub unit and centrifugal clutch assembly – variable ratio models

1 Housing
2 Driven gear
3 Cup bearing – 5 off
4 Idler gear
5 Driving shaft
6 Woodruff key
7 Idler gear
8 Washer
9 Cover assembly
10 Oil seal
11 Nut
12 Shakeproof washer
13 Clutch drum
14 Starting shoe assembly
15 Rubber stop – 2 off
16 Return spring – 2 off
17 Retaining plate
18 Circlip
19 Oil seal
20 Circlip
21 Washer
22 Washer
23 Back plate/hub unit
24 Plug – 6 off
25 Spring – 3 off
26 Spring – 3 off
27 Plain washer – 3 off
28 Pin – 6 off
29 Return spring – 3 off
30 Clutch shoe assembly
31 Needle roller bearing – 2 off
32 Oil seal
33 Pulley spring
34 Outer pulley half
35 Drive belt
36 Inner pulley half
37 Lock washer
38 Securing nut
39 Rear hub unit
40 Assembled clutch and pulley unit
41 Needle roller bearing

Chapter 1 Part 2 Transmission

27 Pedals, chain and freewheel unit: all models

1 The pedals, chain and freewheel unit see comparatively little wear during the life of the machine. The actual components are very similar to those used on bicycles, but enjoy a life of leisure and are completely isolated from the ravages of the elements. It is fairly safe to assume, therefore, that these parts will last almost indefinitely, given an occasional lubrication with aerosol chain lubricant. The jockey wheel, which is fitted to permit chain adjustment, should be set to give about 13 mm ($\frac{1}{2}$ in) movement with the chain at its tightest point. If, for any reason, the freewheel unit fails, it must be unscrewed from the driving shaft and a new unit fitted. It is not practicable to effect a repair to this component. Note that two types of freewheel unit have been used. These are similar in construction but differ in the method of retention. It will be found that the unit is either threaded onto the shaft, and has a large hexagon on the inner face, or is retained by a single central nut at the outer face of the unit.

28 Automatic speed governor: removal, examination and renovation

1 On all variable ratio models, an automatic speed governor is mounted on the crankshaft end in place of the centrifugal clutch. The central retaining nut can be slackened after the tab washer has been knocked back. If necessary, use a strap wrench to hold the unit, or alternatively, wedge a screwdriver blade between the flywheel rotor fins, taking care not to damage them.
2 Slide the unit off the crankshaft end, making sure that it does not pull apart, allowing the rollers to drop out. The driving belt can be removed next, followed by the inner half of the pulley and the distance piece.
3 If the cover is lifted away from the unit, it will be seen that there are five nylon-cased steel rollers which run in grooves in the main body of the unit. Very little in the way of maintenance is required, with the exception of liberally greasing the rollers and tracks each time the unit is disturbed. Reassembly is a straightforward reversal of the removal sequence.

27.1 Use aerosol lubricant on pedal chain

28.2a Knock back tab washer to release securing nut ...

28.2b ... speed governor and spacer

28.3a Remove centre boss and flange

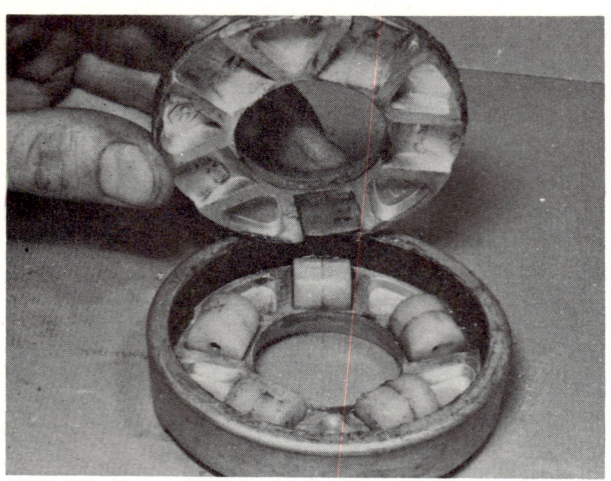

28.3b Lift off cover to expose rollers in housing

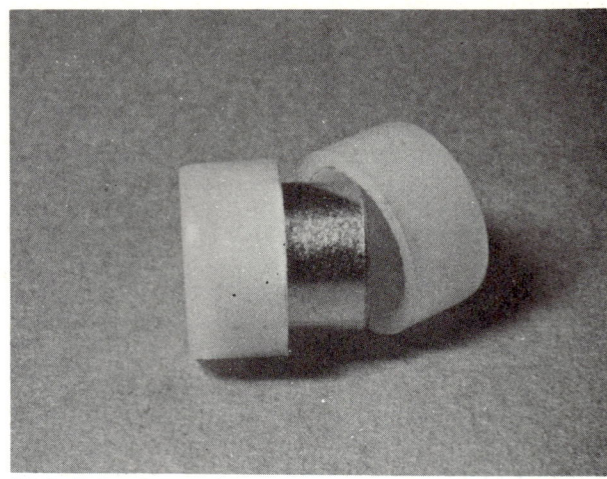

28.3c Rollers have nylon cases

29 Automatic clutch unit: removal, examination and renovation – variable ratio models

1   The automatic centrifugal clutch unit employed on all variable ratio models is similar to that of the single speed machines. It is carried in unit with the split rear pulley on the hub unit input shaft. Reference can be made to Section 2 for details of removal, examination and renovation, noting the following points:

2   The unit is retained by a central nut. With this removed, the drum can be pulled off, complete with the starting shoe assembly. The clutch shoe assembly and spring loaded split pulley can be removed as a unit after the belt has been detached. Alternatively, the assembly can be slid off the shaft and then disengaged from the belt.

3   The pulley assembly can be released after removing the nut which retains it to the clutch centre boss. No maintenance is normally required on this unit. The clutch centre boss is supported on two needle roller bearings. If these become noisy due to wear, they should be driven out of the boss, using a suitable drift, and new ones tapped into position, using a socket as a drift to ensure that they fit squarely. Renew the oil seal as a matter of course each time the bearings are disturbed.

29.2a Release centre nut and remove outer drum

29.2b Note Woodruff key in shaft

29.2c Examine condition of linings

29.2d Starting shoes are retained by circlip and plate

30 Chainwheel, pedals and pedal shaft: all models

1 This assembly is similar in construction to those of bicycles, with the exception of the pedal shaft being supported by two bushes rather than running in bearings. The pedal cranks are retained by cotter pins, which lock the pedal cranks in position in relation to the pedal shaft. To remove the pedal cranks, unscrew the nut until it is flush with the top of the thread. The nut can then be tapped lightly to free the tapered pin. Remove the nut completely and displace the pin to release the pedal. Reassembly is a straightforward reversal of the removal sequence.

2 The pedal shaft can be withdrawn after the left-hand pedal crank has been removed. Normally, it will be sufficient to grease the shaft and bushes occasionally, as they are subject to very little wear. If the bushes do become worn, they can be drifted out of their housings and new bushes tapped into position. The pedals can be removed by unscrewing them from their respective cranks. The right-hand pedal has a conventional right-hand thread, whilst the left-hand pedal has a left-hand thread.

3 As with the other related components, the chainwheel is not likely to show any real degree of wear. Maintenance is confined to keeping the teeth clean and well lubricated, in conjunction with the chain.

31 Drive belt adjustment

1 Provision is made on all fixed ratio models for adjusting the tension of the drive belt. This is accomplished by moving the engine in the frame members to obtain the correct tension. It is important to maintain this setting; an overtight belt will place an excessive loading on the engine main bearings and the rear hub unit bearings, whilst a slack belt will allow slip to occur in the transmission.

2 It is an easy matter to check the tension of the belt if it is suspected of being set incorrectly. This, of course, must be attended to whenever the setting is lost due to a part of the transmission system being disturbed.

3 Remove the left-hand plastic side cover to gain access to the belt. Measure along the top run of the belt a distance of 315 mm (12·4 in) from the rear pulley. Apply a downward loading of 2 kg (4·4 lbs) by suspending weights from the belt or by pulling it down with a spring balance. The belt should deflect by 13·5 – 15·5 mm (0·5 – 0·6 in) when the tension is correct.

4 To adjust the belt tension, slacken the three engine mounting bolts and move the engine backwards or forwards until the correct setting is obtained. Note that the silencer mounting bolt must also be released so that it can move with the engine. Do not forget to tighten all the bolts concerned before refitting the outer cover.

5 Note that no provision is made for belt tension adjustment on variable ratio models. This is because the tension is maintained by the rear spring loaded pulley. If slip does occur, it is because the belt is worn out or contaminated, or the clutch is in need of attention.

30.1a Pedals are retained by cotter pins

30.1b Note spacer on pedal shaft

30.1c Pedal shaft assembly can be withdrawn

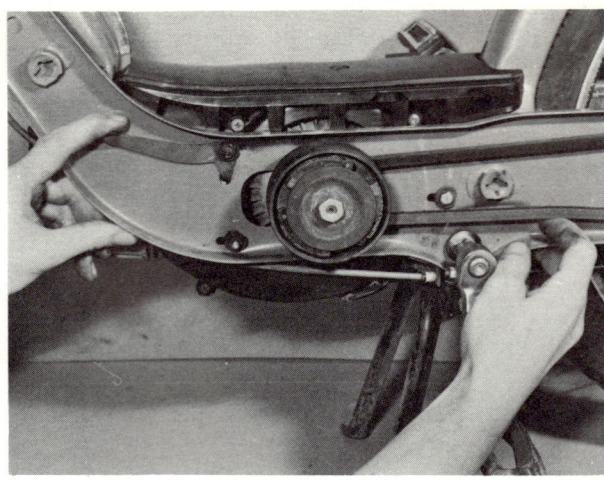

31.4 Lever provides belt tension adjustment

32 Fault diagnosis: engine

Symptom	Cause	Remedy
Engine will not start	Defective spark plug	Remove plug and lay it on cylinder head. Check whether spark occurs when engine is kicked over.
	Dirty or closed contact breaker points	Check condition of points and whether gap is correct.
	Fuel tank empty	Refill.
Engine runs unevenly	Ignition and/or fuel system fault	Check as though engine will not start.
	Blowing cyhead joint	Oil leak should provide evidence. Check for warpage.
	Choked silencer	Remove and clean.
Lack of power	Fault in fuel system	Check system.
	Choked silencer.	See above.
White smoke from exhaust	Engine needs rebore	Rebore and fit oversize piston.
	Tank contains two-stroke petroil of wrong ratio	Drain and refill with petroil at 2%.
Engine overheats	Pre-ignition and/or weak mixture	Check carburettor settings, also grade of plug fitted.

33 Fault diagnosis – transmission

Symptom	Cause	Remedy
Engine fails to turn over when pedalled	Clutch slip	Dismantle and overhaul clutch unit.
	Belt too slack (single speed models)	Check and adjust belt tension.
	Belt worn	Check and renew.
	Starter shoes worn or spring broken	
Renew shoes and/or spring		
Engine runs normally but road speed abnormally slow	Belt worn or slack	See above.
	Speed governor jammed (variable speed models)	Dismantle and overhaul.
	Clutch shoes worn	
Renew		
No drive between engine and rear wheel	Hub unit gears disengaged or broken	Dismantle and overhaul – check whether spring is broken.
No drive between pedals and rear wheel	Freewheel unit broken	Check and renew.

Chapter 2 Fuel system and lubrication

Contents

General description ... 1	Carburettor: dismantling and examination ... 7
Petrol tank: removal (Bravo models only) ... 2	Carburettor: adjusting ... 8
Petrol tank: flushing (All models) ... 3	Air Cleaner: removing, cleaning and replacing ... 9
Petrol tap: removal, dismantling and replacement ... 4	Exhaust system: cleaning ... 10
Petrol feed pipe: examination ... 5	Fault diagnosis: fuel system and lubrication ... 11
Carburettor: removal ... 6	

Specifications

Ciao
Total fuel capacity	2.8 litres (0.62 Imp gall, 0.75 US gall)
Reserve fuel capacity	0.5 litres (0.88 Imp pint, 1.06 US pints)

Bravo
Total fuel capacity	3.0 litres (0.66 Imp gall, 0.80 US gall)
Reserve fuel capacity	0.5 litres (0.88 Imp pint, 1.06 US pint)

Carburettor
Make	Dell'Orto
Type	SHA 12/10
Venturi diameter	10 mm
Main jet	43/100 (49/100 later models)

Engine lubrication By pre-mixed petroil
Mixing ratio	2% or 50:1
	¼ pint oil to 1½ gall petrol
	20 cc oil to 1 litre petrol
Oil grade	SAE 30 self-mixing two-stroke oil

1 General description

The fuel system comprises a petrol tank, from which petrol is gravity fed, via a petrol tap and pipe, to the float chamber of the Dell'Orto carburettor.

For cold starting, a mechanical choke device is fitted. The operating lever protrudes through the side of the top engine cover. When depressed, a blade partially obstructs the airflow through the carburettor, enrichening the mixture entering the engine. As the throttle is opened, a cam on the throttle valve releases the choke mechanism, which then returns to its former position.

Engine lubrication is by petroil-oil carried in solution in the petrol. The incoming mixture is drawn into the crankcase, where it comes into contact with the various moving parts. A proportion of the oil is deposited during the combustion stroke. This system provides lubrication for the big end, small end and main bearings and also for the cylinder walls.

2 Petrol tank: removal (Bravo models only)

1 The petrol tank on Bravo models can be removed for cleaning after releasing the petrol feed pipe from the tap. Prise the pipe off with a screwdriver, having first ensured that the tap is turned off. The tank is retained by a central bolt and two bolts which pass through the lower edge of the tank. Reassembly is by direct reversal of the removal sequence.

3 Petrol tank: flushing (all models)

1 The fuel tank may need flushing out occasionally to remove any accumulated debris which inevitably builds up over the years. This is especially true if water has contaminated the fuel, as this can cause persistent and annoying running problems as it gets drawn into the carburettor. On Bravo models, flushing is best done by removing the tank as detailed in the preceding

Section. On Ciao models the tank is an integral part of the frame, and can best be flushed by removing the fuel tap and flushing through with clean petrol.

4 Petrol tap: removal, dismantling and replacement

1 The fuel taps fitted to both Ciao and Bravo models are similar in construction, differing only in the location of the unit on the two different types of fuel tank. On Ciao models, the tap assembly screws onto a threaded boss on the lower right-hand side of the fuel tank, and can be removed after the side cover has been detached. The Bravo fuel tap, on the other hand, is secured in a similar manner at the front of the petrol tank.
2 The tap can be dismantled for cleaning or examination by unscrewing the hexagon headed gland nut through which the tap spindle passes. Note the position of the components in relation to each other - if they are replaced incorrectly, the aluminium distributor plate will not register correctly. If the tap has been leaking, it is likely that the synthetic rubber seal is worn and requires renewal. Reassembly is a direct reversal of the dismantling sequence.

5 Petrol feed pipe: examination

1 The condition of the plastic petrol feed pipe should be checked periodically. It is likely that the pipe will deteriorate with age, and will require renewal should this occur. Look particularly for signs of splitting or cracking where the pipe pushes over the carburettor and fuel tap stubs. If the pipe has taken a set around one or both of the stubs, it will eventually start to leak. This can be rectified by slicing about $\frac{1}{2}$ in off the pipe and refitting it. For obvious reasons, this cannot be done very often, and ultimately it will be necessary to renew the pipe.

6 Carburettor: removal

1 The carburettor on all models is retained by a clip to a stub at the rear of the crankcase. The method of removal is similar in all cases, but it should be noted that on fixed ratio models, it may be advisable to slacken the engine mounting bolts, detach the drive belt, and then slide the engine fully forward to give the best possible access to the carburettor.
2 Slacken the four top cover retaining screws and lift the cover away (Bravo models). On Ciao models, release the central fixing screw to release the cover. Remove the right-hand side cover. The petrol pipe should be detached, either by prising it off the stubs to which it is attached, or by releasing the banjo union from the carburettor body. Make sure that the fuel tap is in the off position first!
3 Slacken the two screws which retain the mixing chamber top. This can then be pulled away, complete with the valve assembly and return spring. These components need not be removed from the throttle cable at this stage.
4 Slacken the clamp which retains the air cleaner unit to the carburettor body, and pull it clear. On Bravo models, it will be necessary to twist the air cleaner unit to manoeuvre it out from between the engine side members and the frame tube.
5 The clamp retaining the carburettor to the crankcase can now be slackened, and the instrument pulled back, away from its mounting stub, and then lifted clear of the frame. Note that on Ciao models it is advisable to remove the top cover mounting bracket to give better access. The latter is secured by two set screws to the right-hand side member.

7 Carburettor: dismantling and examination

1 The type of carburettor used on both Ciao and Bravo models is very simple in operation and construction. In consequence, it is very easy to dismantle and clean. The instrument has only one jet and one adjustment screw.
2 Remove the two screws which retain the float chamber, and lift it away, ensuring that the O ring is not damaged. Displace the float pivot pin which retains the float to its mounting lugs. Lift the float away, taking care that the float needle is not lost. The float chamber and needle valve assembly should be cleaned carefully, noting any foreign matter or water which may have been responsible for erratic running. Check that the seating faces of the float needle and valve seat are in good condition, otherwise flooding may result.
3 The main jet can be unscrewed from the pillar in which it seats, for cleaning and examination. Use compressed air to clear an obstruction in the orifice, preferably using a high pressure air line, or alternatively, a foot pump or similar. As a last resort, a fine *nylon* bristle can be used, but on no account use wire or a pin as this will damage the precision drilling of the jet.
4 Examine the throttle valve for signs of wear by disengaging it from the throttle cable, and placing it in position in the carburettor body. Any excessive play between the two will allow air to leak past, upsetting the mixture and causing erratic running. The only remedy is renewal of the part(s) which are affected.
5 Reassemble the carburettor in the reverse order of the dismantling sequence, ensuring that everything is kept scrupulously clean. Before fitting the throttle valve, it should be lubricated with light machine oil.
6 Before use, check for leaks, and check that the settings are in order as described in the following Section. If dismantling was necessary because of contaminated fuel, remove and flush the fuel tank, and clean the filter in the tap. Water can cause persistent and erratic running faults, and can form by condensation in the tank.

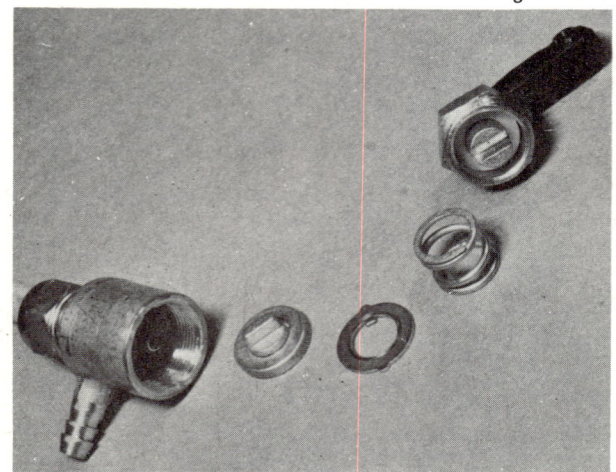

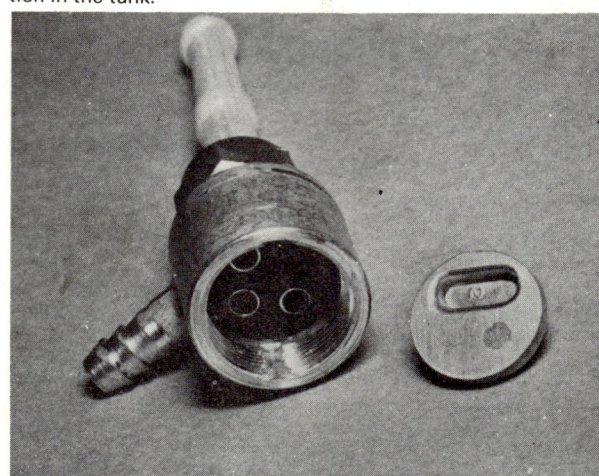

4.2a Unscrew nut to release fuel tap components

4.2b Petrol is channelled through slot in aluminium block

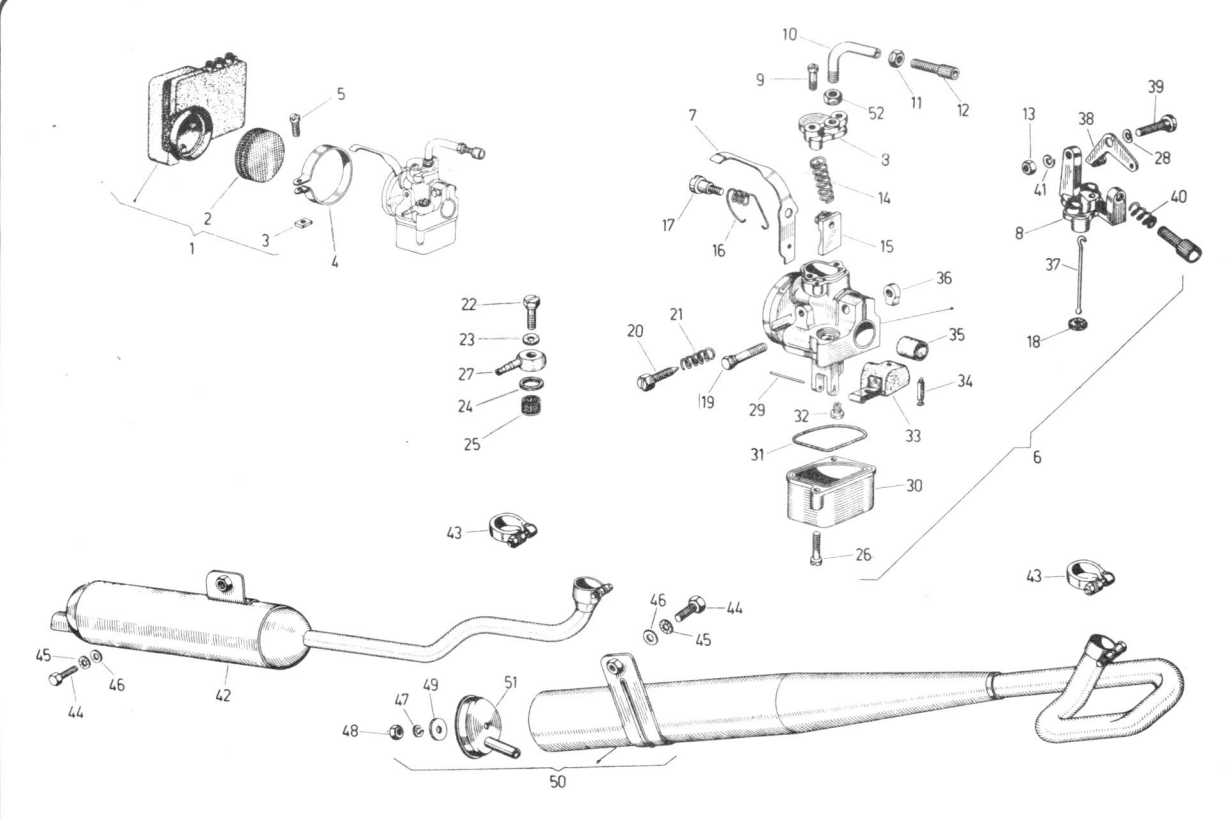

Fig. 2.1 Carburettor and exhaust system

1 Air filter assembly	14 Return spring	27 Banjo union	40 Spring
2 Gauze element	15 Throttle valve	28 Wave washer	41 Spring washer
3 Threaded plate	16 Return spring	29 Float pivot pin	42 Exhaust system
4 Mounting plate	17 Shouldered screw	30 Float chamber	43 Clamp
5 Screw	18 Damping washer	31 Seal	44 Bolt
6 Carburettor assembly	19 Pinch bolt	32 Main jet	45 Shakeproof washer
7 Cold start control	20 Throttle stop screw	33 Float	46 Plain washer
8 Carburettor top	21 Spring	34 Float needle	47 Spring washer
9 Screw – 2 off	22 Banjo bolt	35 Insulating bush	48 Nut
10 Cable guide	23 Fibre washer	36 Threaded block	49 Plain washer
11 Lock nut	24 Fibre washer	37 Throttle valve guide	50 Exhaust system
12 Adjuster	25 Fuel filter	38 Pivot lever	51 Tail pipe/cover
13 Nut	26 Screw – 2 off	39 Shouldered screw	

6.2 Release pipe at banjo union – Note filter element A and throttle stop screw B

6.3 Throttle top is retained by two screws (arrowed)

7.2a Release float chamber screws, and lift away

7.2b Displace pivot pin to release float and needle

7.3 Main jet screws into holder

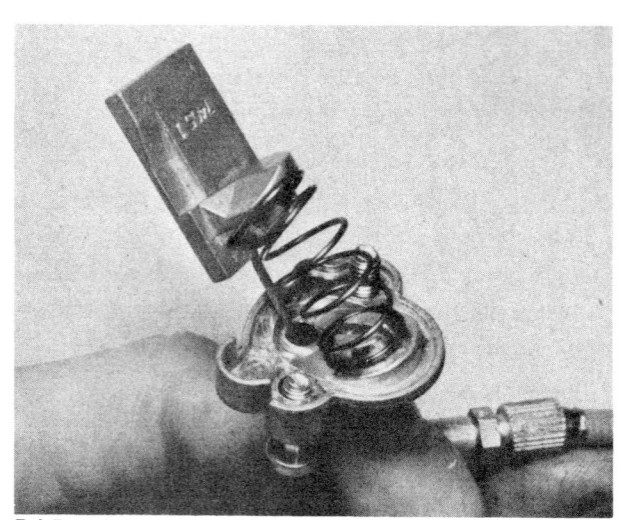

7.4 Examine throttle valve components for wear

7.5a Refit float and pivot pin. Note throttle stop screw (arrowed)

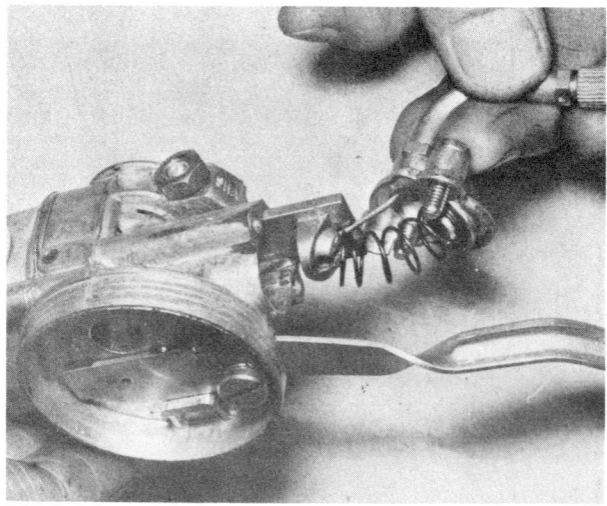

7.5b Lubricate throttle valve before reassembly

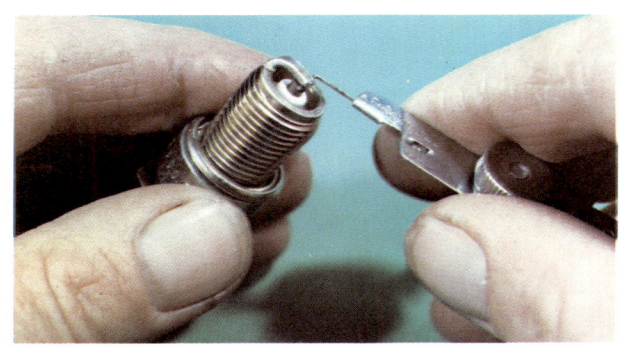

Spark plug maintenance: Checking plug gap with feeler gauges

Altering the plug gap. Note use of correct tool

Spark plug conditions: A brown, tan or grey firing end is indicative of correct engine running conditions and the selection of the appropriate heat rating plug

White deposits have accumulated from excessive amounts of oil in the combustion chamber or through the use of low quality oil. Remove deposits or a hot spot may form

Black sooty deposits indicate an over-rich fuel/air mixture, or a malfunctioning ignition system. If no improvement is obtained, try one grade hotter plug

Wet, oily carbon deposits form an electrical leakage path along the insulator nose, resulting in a misfire. The cause may be a badly worn engine or a malfunctioning ignition system

A blistered white insulator or melted electrode indicates over-advanced ignition timing or a malfunctioning cooling system. If correction does not prove effective, try a colder grade plug

A worn spark plug not only wastes fuel but also overloads the whole ignition system because the increased gap requires higher voltage to initiate the spark. This condition can also affect air pollution

7.5c Cold-start lever obstructs airflow when in use

8 Carburettor: adjusting

1 The carburettor is provided with one adjustment; namely, tickover speed. The other settings are pre-determined by the manufacturer and cannot be altered. A throttle stop screw is provided so that a reliable and even tickover speed can be maintained.

2 Before any adjustment is made, check, and if necessary, reset the amount of free play in the throttle cable. The adjuster is incorporated at the carburettor end of the cable for this purpose. When set correctly there should be about $\frac{1}{8}$ in free play before the throttle valve commences to move.

3 Start and run the engine until normal operating temperature is reached. Place the machine on its centre stand and allow the engine to idle. The manufacturers recommend that the setting is correct when the tickover speed is consistent, and the rear wheel is not turning due to the centrifugal clutch being partially engaged. The adjusting screw can be reached via a hole in the right-hand side member. Turn the screw clockwise to raise the tickover speed and vice versa. If any significant adjustment has to be made, ensure that the cable free play is rechecked before replacing the outer covers.

9 Air cleaner: removing, cleaning and replacing

1 The air cleaner consists of a baffled plastic case containing a gauze filter element. It is retained to the carburettor by a clamp fixing. Access to the air cleaner is gained after releasing the top plastic cover. Slacken the clamp screw and lift the air cleaner clear of the carburettor. Note that on fixed ratio machines it may prove necessary to move the engine unit fully forwards as described in the preceding Section, thus giving better clearance for removal.

2 Lift the gauze element out of the recess in the front of the filter case. Wash both components in petrol to remove any traces of dust or other foreign matter. Before refitting the air cleaner, wipe the gauze with a little light oil to provide a better filtering effect. The oil will trap any small particles which might otherwise enter the engine.

10 Exhaust system: cleaning

1 The exhaust system is a one-piece welded unit comprising an exhaust pipe and silencer unit. It should be detached for cleaning after slackening the clamp at the exhaust port and removing the single silencer retaining bolt which passes through the side member.

2 The areas most likely to require attention are the exhaust pipe and tailpipe, which will tend to become choked if not kept clear. A two-stroke engine is very susceptible to this fault, which is caused by the oily nature of the exhaust gases. As the sludge builds up back pressure will increase with a resulting fall off in performance.

3 If the build up of carbon and oil is not too great, a wash with a petrol/paraffin mix will probably suffice as the cleaning medium. Otherwise more drastic action will be necessary such as the application of a blowlamp flame to burn away the accumulated deposits.

4 If the painted finish of the system becomes damaged, it should be attended to quickly, or the system will rapidly become corroded and need renewal. Remove any surface rusting, and paint it with a proprietary high-temperature enamel. One of the numerous rust inhibiting fluids may also be used to good effect, if rusting has already taken place.

9.2 Clean filter element in petrol to remove dust

10.3 Remove exhaust system for cleaning, if badly choked

Chapter 2 Fuel system and lubrication

11 Fault diagnosis – fuel system

Symptom	Cause	Remedy
Excessive fuel consumption	Air cleaner choked or restricted	Clean or renew element.
	Fuel leaking from carburettor	Check all unions and gaskets.
	Badly worn or distorted carburettor	Renew.
Idling speed too high	Throttle stop screw in too far	Adjust screw.
	Carburettor top loose	Tighten.
Engine sluggish. Does not respond to throttle	Back pressure in silencer	Check and clean if necessary.
Engine dies after running for a short while	Dirt or water in carburettor	Remove and clean.
General lack of performance	Weak mixture; float needle sticking in seat	Remove float chamber and check needle seating.
	Air leak at carburettor or leaking crankcase seals	Check for air leaks or worn seals.

12 Fault diagnosis – lubrication system

Symptom	Cause	Remedy
White smoke from exhaust	Too much oil in fuel	Drain and refill with correct mixture.
Engine runs hot and gets sluggish when warm	Too little oil in fuel	See above.

Chapter 3　Ignition system

Contents

General description 1	Condenser: checking and renewal 5
Ignition timing: checking and setting 2	Ignition coil: removal and replacement 6
Contact breaker points: adjustment 3	High tension (sparking plug) lead: examination 7
Contact breaker assembly: removal and replacement ... 4	Fault diagnosis: Ignition system 8

Specifications

Ignition system	Flywheel magneto with external ignition coil
Contact breaker gap	0.35–0.45 mm (0.013–0.017 in)
Sparking plug gap	0.60 mm (0.24 in)
Sparking plug type	Marelli CW4N AT Bosch W95T1 Motorcraft AE 4

1　General description

1　The spark necessary to ignite the petrol/air mixture in the combustion chamber is derived from a crankshaft-mounted flywheel magneto. A contact breaker assembly, contained in the flywheel magneto, determines the exact point at which the spark occurs. As the points separate, the low tension circuit is interrupted, and a high tension voltage is developed in the ignition coil. This passes across the electrodes of the sparking plug, igniting the compressed mixture as it jumps across the air gap between the two electrodes.

The flywheel magneto also produces power for the electrical system of the vehicle. This aspect of its function is covered in Chapter 6 of this Manual.

2　Ignition timing: checking and setting

1　The ignition timing on all Ciao and Bravo models is preset by the manufacturer on assembly. As no provision is made for adjustment, it is normally impossible for the setting to be incorrect. The exception to the above rule is when the contact breaker points or fibre heel become worn, in which case the accuracy of the timing will be affected. In this instance, renewal of the contact breaker assembly will restore the ignition timing to its correct setting.

2　The factor which governs the spark advance setting during manufacture is the position of the Woodruff key slot in the mainshaft. For this reason, it is important that the correct crankshaft for any particular type of engine is used. Make sure that this condition is complied with if a new crankshaft assembly is to be fitted. A Vespa Service Agent will be able to supply the correct component, providing that the engine number is quoted when ordering it.

3　Contact breaker points: adjustment

1　Remove the left-hand outer cover, and turn the engine over until the rubber plug in the flywheel is visible through the gap in the engine side member. Prise out the plug and rotate the engine slowly to and fro until the contact breaker points are fully opened. Check that the point faces are not excessively burnt or pitted. If they are, remove and renew the assembly as described in Section 4 of this Chapter.

2　If the contacts are in good condition, measure the gap using a feeler gauge. A 0.4 mm (0.015 in) gauge should be a light sliding fit (the points **must** be within the range: 0.35 – 0.45 mm (0.013 –0.017 in). Should they require adjustment, slacken the securing screw *just* enough to permit the fixed contact to be moved, using a small screwdriver. Tighten the securing screw and then recheck the gap. If the gap is set correctly, it may be assumed that the ignition timing is also accurate. Do not omit to replace the rubber plug, to prevent the ingress of water during use.

Chapter 3 Ignition system

3.2 Check contact breaker gap using a feeler gauge

4 Contact breaker assembly: removal and replacement

1 If examination of the point faces has shown them to be pitted or worn, it will be necessary to remove them for renewal. This entails the removal of the engine from the frame, and the extraction of the flywheel rotor, using the correct service tool. This is not as daunting as it might at first appear. Refer to Chapter 1, Part 1, Sections 4 and 7 for full details.

2 With the flywheel rotor removed, release the screw which retains the contact breaker leads, noting the **exact** order in which they are fitted. (Make a sketch of the leads to avoid any confusion during reassembly). Remove the fixed contact retaining screw, and lift the assembly away from the casting.

3 **On no account** reface the contact breaker points in the hope of re-using them. If this is done the ignition timing will be affected. When fitting the new contact breaker assembly, apply a few drops of light oil to the lubricating wick to offset wear of the fibre heel.

5 Condenser: checking and renewal

1 A condenser is included in the contact breaker circuitry to prevent arcing across the contact breaker points as they separate. It is connected in parallel with the points and if a fault develops, ignition failure will occur.

2 If the engine is difficult to start, or if misfiring occurs, it is possible that the condenser is at fault. This symptom will be underlined if the contact breaker points have eroded quickly and have a burnt or blackened appearance. A replacement condenser is inexpensive, and if suspect in any way, the condenser should be renewed. It may even be considered worthwhile renewing the condenser each time the contact breakers are renewed, in view of the need to remove the engine and flywheel rotor to gain access to the components.

3 Remove the engine from the frame, and extract the flywheel rotor as described in the preceding Section, if this has not already been done. The condenser will be found attached to the crankcase, opposite the contact breaker assembly. The method of renewal of the condenser is self-evident. Check that the leads are correctly assembled on the contact breaker assembly.

6 Ignition coil: removal and replacement

1 The ignition coil will be found bolted to the underside of the engine, or on some models, to the right-hand side member. This unit rarely fails in service, and requires no attention during the normal life of the machine.

If, however, it does fail, complete loss of ignition will result. If the fault cannot be attributed to the sparking plug, HT lead, contact breaker assembly or condenser, the unit should be detached and entrusted to a Vespa Service Agent or Auto-Electrician for testing. The coil cannot be tested without specialist equipment, and cannot be repaired at home. In event of failure it is best to renew the unit.

4.2 A: Lead securing screw. B: Fixed contact securing screw

6.1 Coil is mounted here or on underside of engine

7 High tension (sparking plug) lead: examination

1 Erratic running faults and problems with the engine suddenly cutting out in wet weather can often be attributed to leakage from the high tension lead and sparking plug cap. If this fault is present, it will often be possible to see tiny sparks around the lead and cap at night. One cause of this problem is the accumulation of mud and road grime around the lead, and the first thing to check is that the lead and cap are clean. It is often possible to cure the problem by cleaning the components and sealing them with an aerosol ignition sealer, which will leave an insulating coating on both components.

2 Water dispersant sprays are also highly recommended where the system has become swamped with water. Both these products are easily obtainable at most garages and accessory shops. Occasionally, the suppressor cap or the lead itself may break down internally. If this is suspected, the components should be renewed. It is recommended that the renewal of the lead is entrusted to an Auto-Electrician who will have the expertise to solder on a new lead without damaging the coil windings.

8 Sparking plug: checking and resetting the gap

1 The Ciao and Bravo models are fitted with a Marelli CW4N AT or a Bosch W95TI sparking plug as standard equipment. The electrode gap should be maintained at 0.6 mm (0.024 in). Certain operating conditions may indicate a change in sparking plug grade, although the type recommended by the manufacturer will usually give the best, all round service. The use of anything other than the recommended grade may result in a holed piston.

2 Check the gap of the plug points during every three monthly or two thousand mile service. To reset the gap, bend the outer electrode to bring it closer to the centre electrode and check that the correct feeler gauge can be inserted. Never bend the central electrode or the insulator will crack, causing engine damage if the particles fall in whilst the engine is running.

3 With some experience, the condition of the sparking plug electrode and insulator can be used as a reliable guide to engine operating conditions. See accompanying diagram.

4 Beware of overtightening the sparking plug otherwise there is risk of stripping the threads from the aluminium alloy cylinder head. The plug should be sufficiently tight to sit firmly on its copper sealing washer, and no more. Use a spanner which is a good fit to prevent the spanner slipping and breaking the insulator.

5 If the threads in the cylinder head strip as a result of overtightening the sparking plug, it is possible to reclaim the head by use of a Helicoil thread insert. This is a cheap and convenient method of replacing the threads; most motorcycles dealers operate a service of this kind.

6 Make sure that the plug insulating cap is a good fit and has its rubber seal. It should also be kept clean to prevent tracking. The cap contains the suppressor that eliminates both radio and television interference.

9 Fault diagnosis – ignition system

Symptom	Cause	Remedy
Engine will not start	No spark at plug	Faulty ignition lead – check connections and insulation. Check plug cap.
	Weak spark at plug	Dirty contact breaker points require renewal. Contact breaker gap has closed up. Reset.
	Faulty HT coil	Renew.
Engine starts, but runs erratically	Intermittent or weak spark	Check condition of spark plug. If no improvement check whether points are arcing. If so renew condenser.
	Plug lead insulation breaking down	Check for breaks in outer covering, especially near frame.
Engine difficult to start and runs sluggishly. Overheats	Faulty contact breaker points	Check condition of contact breaker and renew if worn.

Chapter 4 Frame and forks

Contents

General description	1
Rigid front forks: removal, examination and renewal	2
Leading link front forks: removal and renovation	3
Hydraulic front forks: removal and reassembly	4
Oil-bath type telescopic forks: removing the legs from the yokes	5
Front fork legs: dismantling, renovation and reassembly (all-steel type)	6
Front fork legs – dismantling, renovation and reassembly (oil-bath type)	7
Frame: examination and renovation	8
Swinging arm rear suspension: examination and renovation	9
Rear suspension units: removal and replacement	10
Centre stand: examination	11
Saddle: adjustment	12
Speedometer head: removal and replacement	13
Speedometer drive cable: examination and lubrication	14
Cleaning the machine	15
Fault diagnosis: Frame and forks	16

Specifications

Front forks
Types ... Rigid, undamped leading-link or hydraulic
Oil capacity (hydraulic forks) ... to level plug
Oil grade (hydraulic forks) ... SAE 20W engine oil or fork oil

Rear suspension
Types ... Rigid frame, sprung saddle, (Ciao)
Swinging arm (Bravo)

Suspension units ... Coil spring type

---

16 Fault diagnosis – frame and forks

1 General description

The Vespa Ciao and Bravo mopeds employ a welded tubular steel frame, having the engine and rear wheel supported by pressed steel side members. Four types of front fork are employed: The rigid forks are of welded pressed steel construction and are similar to those used on bicycles. The leading link type forks are basically similar to the rigid types, but have short alloy links pivoted at the lower end of the fork blades and supported by short coil springs. The hydraulic forks are of the single or double spring type, having internal coil springs.

On rigid framed models, the rear wheel is mounted directly in the frame. To provide some measure of comfort for the rider, a sprung cantilever saddle is employed to absorb road shocks. Bravo models employ a form of swinging arm rear suspension in which the pressed steel side members are attached to the main frame tube by way of a welded lug above the engine unit. The complete assembly pivots about this lug, being supported by telescopic rear suspension units.

2 Rigid front forks: removal, examination and renewal

1 As there are no moving parts involved in these forks, the only normal reason for removal is in the event of accident damage where the forks require renewal. It is not a practicable or safe proposition to attempt to straighten the fork blades if they should become twisted or bent. The forks can be removed as described below:

2 Place the machine securely on its centre stand and arrange it so that the front wheel is raised clear of the ground. Release the front brake cable and speedometer drive cable. Slacken the wheel spindle nuts at each side and drop the wheel free from the fork ends. Remove the wheel and place it to one side.

3 Slacken the long chromium plated steering column bolt by about ½ in. Using a soft-faced mallet, tap the bolt downwards to drive the cone out of the bottom of the stem, freeing the steering column. The handlebars can now be pulled upwards and free of the steering head, complete with all cables and controls.

4 Slacken the steering head locknut and top cup. With these removed, the fork assembly can be lowered clear of the frame, complete with the caged ball headraces (2 off). The front mudguard can be removed after releasing the three screws which retain it.

5 The forks can be refitted in the reverse order of that described for dismantling, ensuring that the upper and lower head races are greased generously. When fitting the top cup, take great care not to overtighten it. Tighten it *just* enough to remove any trace of play then fit and tighten the locknut. Note that a loading of several tons can be placed on the bearings if this precaution is not observed. Make sure that the handlebars are set at the correct position before tightening the steering column bolt.

Chapter 4 Frame and forks

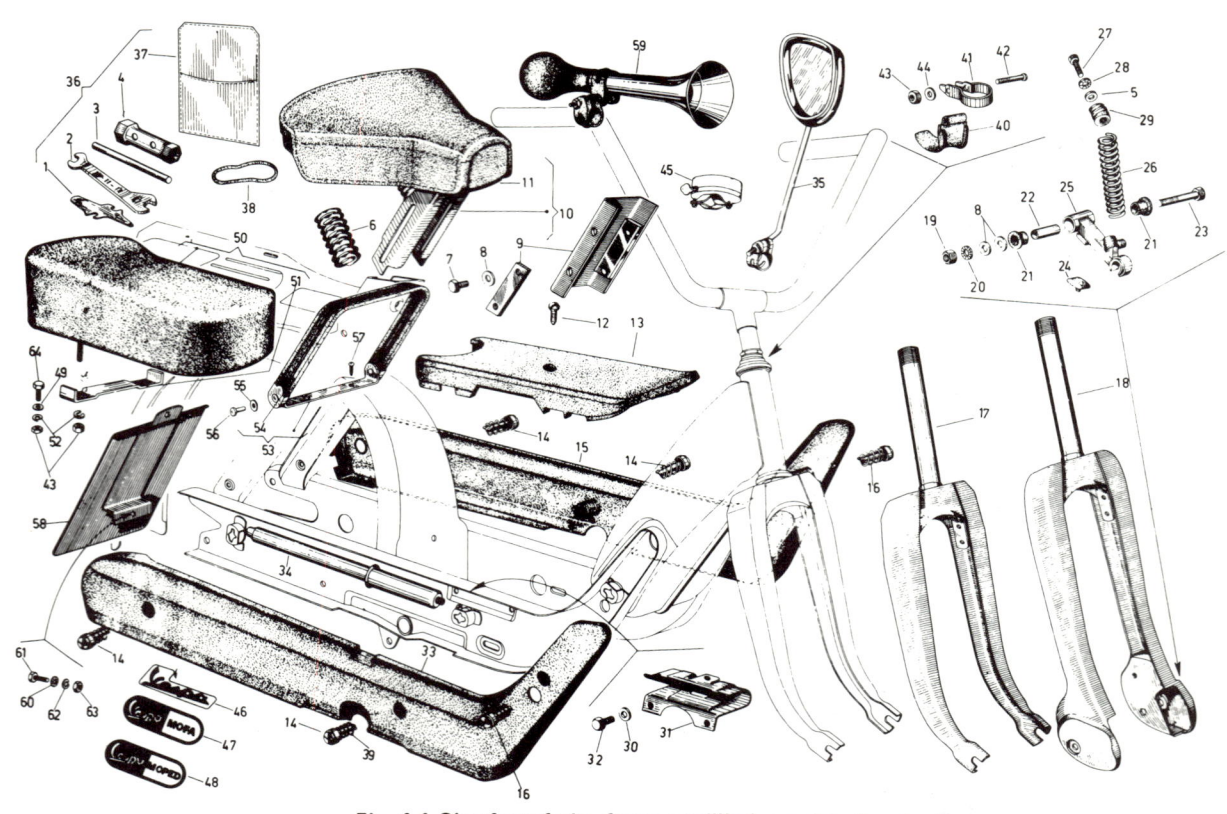

Fig. 4.1 Ciao front forks, frame ancilliaries and optional extras

1. Screwdriver/wrench
2. Open ended spanner
3. Tommy bar
4. Box spanner
5. Plain washer – 2 off
6. Saddle spring
7. Saddle mounting bolt – 2 off
8. Plain washer – 2 off
9. Adjusting plate – 2 off
10. Saddle assembly (not SC models)
11. Saddle cover
12. Top cover securing screw
13. Top cover
14. Long fastener – 4 off
15. Left-hand side cover
16. Short fastener
17. Rigid fork complete
18. Leading link fork bare
19. Nut – 2 off
20. Lock washer – 2 off
21. Bush – 4 off
22. Spacer – 2 off
23. Link pivot bolt – 2 off
24. Rubber damping block – 2 off
25. Link
26. Spring – 2 off
27. Bolt – 2 off
28. Shakeproof washer – 2 off
29. Retaining block – 2 off
30. Plain washer – 2 off
31. Top cover bracket
32. Bolt – 2 off
33. Right-hand outer cover
34. Tyre inflator
35. Mirror
36. Tool kit assembly
37. Bag for tool kit
38. Spring strap
39. Spring – 6 off
40. Hook
41. Clip
42. Screw
43. Nut – 3 off
44. Washer
45. Bell
46. Badge – 2 off
47. Badge – 2 off
48. Badge – 2 off
49. Plain washer – 2 off
50. Optional passenger seat
51. Seat bracket
52. Spring washer – 2 off
53. Grab handle bracket
54. Grab handle
55. Washer – 2 off
56. Pivot pin – 2 off
57. Screw – 2 off
58. Rear number plate
59. Bulb horn
60. Shakeproof washer
61. Bolt
62. Spring washer – 3 off
63. Nut – 3 off
64. Bolt – 2 off

3 Leading link front forks: removal and renovation

1 The leading link type of front fork can be dealt with in much the same way as that described in the preceding Section. The links and springs themselves can be dismantled individually, after the front wheel has been removed. There is no need to remove the complete forks to dismantle these components.

2 The links and springs are contained in small welded boxes which form the lower end of the fork blades. Each link is supported on a bolt which passes through the box section. The links are shaped so that the spring will screw onto them. A corresponding cast boss retains the other end of the spring to the fork.

3 The spring end is retained either at the top or to the rear of the box section. (The links used on the former type have a grease nipple fitted). This does not affect the method of removal or reassembly, but it should be noted that the two types are not interchangeable.

4 Remove the nut (or bolt, on early models), shakeproof washer and plain washer which retains the end of each spring. Remove the nut and washer on which each link pivots. The bolt(s) can now be displaced, using a drift to tap it (them) free, and the link(s) and spring(s) removed.

5 Examine the bushes in the links for wear. If badly worn, play in the suspension will allow wheel wobble to develop. This can be checked by passing the pivot bolt through the bushes and feeling for any discernible amount of sloppiness. If worn, the distance piece should be driven out of the link, followed by the two headed bushes. Renew the bushes, distance pieces and pivot bolts as complete assemblies. Make sure that **both** units are reconditioned at the same time.

6 Grease the moving parts of the assembly prior to refitting them into the forks. Fit the spring anchor nut or bolt first, followed by the pivot bolt. It may prove necessary to lever the link into position against spring pressure, and an assistant will be invaluable at this stage.

Chapter 4 Frame and forks

3.4a Springs are retained by bolt or nut, and washers

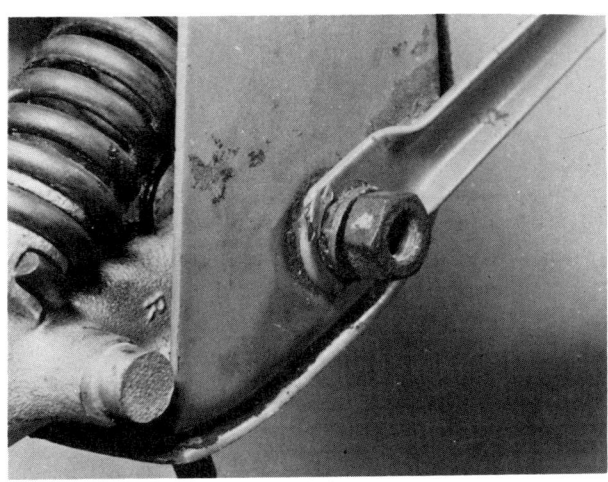

3.4b Displace pivot bolt to free link

3.4c Assembly can now be withdrawn from fork

3.5 Examine bushes and link for signs of wear

4 Hydraulic front forks: removal and reassembly

1 Two types of telescopic front have been employed on Bravo models. The simpler of the two types is basically a helical coil spring contained within two telescopic tubes. The second of the two types has a slightly more sophisticated construction using steel stanchions, light alloy lower legs and a double spring arrangement. This type also contains an oil bath in the lower leg.

2 The all-steel type of fork can be removed only as an assembly, whereas with the oil-bath type, it is possible to remove the individual legs, leaving the steering head assembly undisturbed. (Refer to Section 5 of this Chapter). Before any dismantling work is undertaken, ensure that the machine is securely supported with its front wheel clear of the ground.

3 Detach the speedometer drive cable and the front brake cable from their mounting points on the hub. Slacken the wheel spindle nuts at each side, to allow the wheel to be lowered clear of the forks. It is advisable, though not essential at this stage, to remove the front mudguard from the forks, to avoid any risk of damage to the paint finish. It is retained by two nuts and bolts passing through the lugs on each lower fork leg.

4 To free the complete fork assembly, it will be necessary to release the steering head assembly and the top yoke. The latter can be left in position, together with the headlamp unit and handlebars.

5 Separate the two halves of the headlamp shell by releasing the securing screws on the underside of the unit. Lift the top of the unit clear but do not disconnect it from the speedometer cable and wiring.. Remove the two bolts which pass through the top yoke into the top of the fork stanchions. Remove the nuts from the two U bolts which retain the handlebars, and lift the handlebars away, the support pillars, the U bolts and the flat locking plate.

6 Remove the steering column nut, followed by the upper steering head cup, As the latter is removed, the fork assembly can be lowered clear of the headstock and removed for further dismantling. The handlebars, complete with controls, and the headlamp unit, can be left in situ to await reassembly.

7 Reassembly is a direct reversal of the dismantling process, noting that the caged steering head bearings should be greased generously prior to reassembly. When fitting the top cup, it should be tightened *just* enough to eliminate any play. On no account should this be overtightened, as the excess loading placed on the bearings, which can be several tons, will quickly wear them, and may even crack the balls or races. Refit the steering head nut, noting the flat security plate which prevents this from working loose.

4.3 Mudguard is secured by four studs and nuts

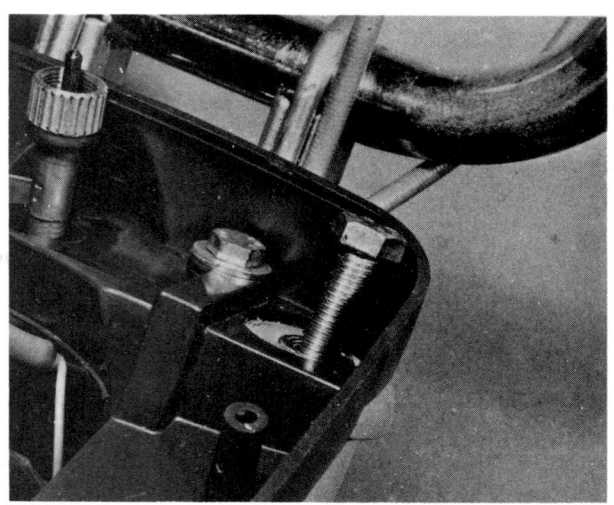

4.5 Remove fork top bolt, located inside headlamp unit

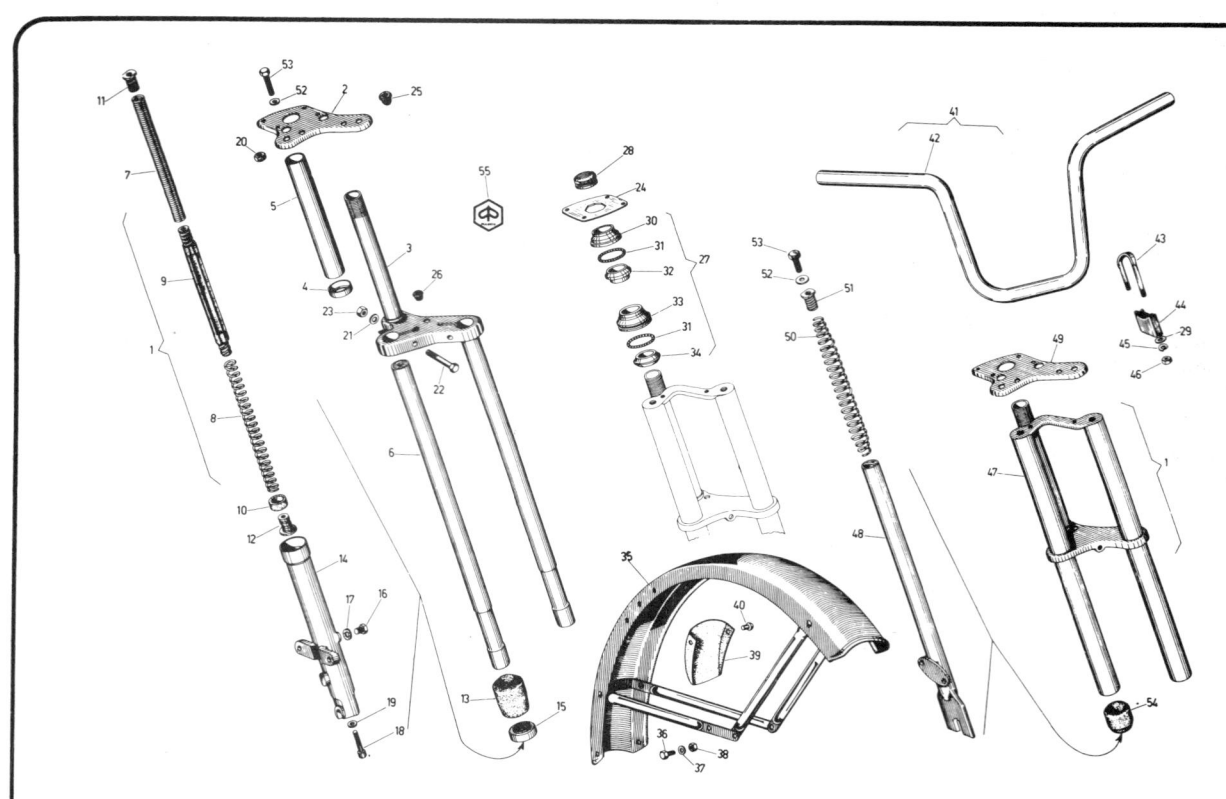

Fig. 4.2 Telescopic front forks

1. Fork/spring assembly
2. Top yoke
3. Lower yoke and steering column
4. Cup – 2 off
5. Shroud
6. Stanchion – 2 off
7. Upper spring – 2 off
8. Lower spring – 2 off
9. Distance piece – 2 off
10. Damping block – 2 off
11. Threaded block – 2 off
12. Lower threaded block – 2 off
13. Gaiter
14. Lower fork leg – 2 off
15. Oil seal – 2 off
16. Drain plug – 2 off
17. Fibre washer – 2 off
18. Damper rod bolt – 2 off
19. Copper washer – 2 off
20. Nut – 2 off
21. Plain washer – 4 off
22. Pinch bolt – 2 off
23. Nut – 2 off
24. Locking plate
25. Grommet
26. Grommet
27. Steering head bearing assembly
28. Steering column top nut
29. Plain washer – 4 off
30. Upper cup
31. Steering head race – 2 off
32. Upper cone
33. Lower cup
34. Lower cone
35. Front mudguard
36. Bolt – 4 off
37. Plain washer – 4 off
38. Nut – 4 off
39. Mudflap
40. Rivet – 2 off
41. Handlebar assembly
42. Handlebar
43. U bolt – 2 off
44. Support bracket – 2 off
45. Spring washer – 4 off
46. Nut – 4 off
47. Upper fork assembly (all-steel)
48. Stanchion
49. Top yoke
50. Helical fork spring – 2 off
51. Threaded block – 2 off
52. Plain washer – 2 off
53. Bolt – 2 off
54. Gaiter
55. Badge

5 Oil-bath telescopic forks: removing the legs from the yokes

1 The oil-bath type fork legs can be removed from the machine individually, if required, in which case the steering head need not be disturbed. Follow the instructions given in Section 4 from paragraphs 3 to 5 inclusive, noting that the handlebars need not be disturbed.
2 The fork legs are now retained only by the clamp bolts which pass through the bottom. When these are released, it will be possible to draw the legs downwards and clear of the yokes. The shrouds can be left in position.
3 Note that during reassembly, it may prove difficult to get the fork leg back into position. If this is the case, obtain a long bolt which will screw into the thread at the top of the leg. This can then be passed down through the shrouds, screwed into the fork leg, and used to draw it up into position.

6 Front fork legs: dismantling, renovation and reassembly (rigid type)

1 Remove the two top bolts, releasing the lower legs complete with springs. The helical spring engages in a thread in the top of the lower leg, and can be unscrewed. A corresponding threaded boss is fitted to the upper end of the spring.
2 Little can be done by way of maintenance to this assembly, and if accident damage has been sustained, the damaged parts should be renewed. When refitting the springs, they should be liberally coated in heavy grease to prevent them chattering in use.

7 Front fork legs: dismantling, renovation and reassembly (oil-bath type)

1 Invert the fork legs and allow the oil to drain into a container of about 2 pints capacity. It is recommended that one fork leg be dealt with at a time, to avoid any risk of parts becoming interchanged. Wrap the lower leg in rag and clamp it firmly in a vice. **Do not** overtighten the vice or the lower leg may be crushed. Using an Allen key, slacken and remove the socket screw which passes up through the bottom of the fork leg. This will release the spring retaining block.
2 Slide the dust seal off the lower leg and draw the two halves of the fork assembly apart. It will be noted that each leg contains two springs, with a plastic distance piece between them. This component screws into the ends of the springs, as do the upper and lower retaining blocks.
3 Examine the various components for signs of wear or damage. The stanchion should be a light sliding fit in the lower leg, without having any discernible side to side play. Check the straightness of the stanchion by rolling it on a flat surface. It may be possible to correct slight bending by straightening the stanchion in a press. Normally, however, it is best to renew the component in view of the risk of weakness. Reassembly is a direct reversal of the dismantling process. Note that each leg should be fitted to the filler/level hole in the lower leg with SAE 20W oil or one of the proprietary fork oils. This is best done after the forks are installed in the frame.

8 Frame: examination and renovation

1 The frame is unlikely to require attention unless it is damaged as the result of an accident. In many cases, replacement of the frame is the only satisfactory course of action, if it is badly out of alignment. Comparatively few frame repair specialists have the necessary mandrels and jigs essential for the accurate re-setting of the frame and, even then there is no means of assessing to what extent the frame may have been overstressed such that a later fatigue failure may occur.
2 After a machine has covered an extensive mileage, it is advisable to keep a close watch for signs of cracking or splitting at any of the welded joints. Rust can cause weakness at these joints particularly if they are unpainted. Minor repairs can be effected by welding or brazing, depending on the extent of the damage found.
3 A frame out of alignment will cause handling problems and may even promote 'speed wobbles' in a particular speed range. If misalignment is suspected as the result of an accident, it will be necessary to strip the machine so that the frame can be checked, and if needs be, renewed.

9 Swinging arm rear suspension: examination and renovation

1 On Bravo models, a form of swinging arm rear suspension is employed, in which the side members carrying the engine, rear wheel and transmission are pivoted at a point just above the engine unit. It is possible to remove this arrangement as a sub-assembly, but there is normally no reason for this to be done.
2 After a period of extended use, it is likely that the bushes on which the sub-assembly pivots will wear, allowing side play to develop. This will eventually give rise to a noticeable twitch, when cornering. It is important that any trace of play is eliminated. Its development is somewhat insidious and can lead to the machine becoming dangerous without the owner realising the problem. Play can be checked by pushing the assembly from side to side. Any discernible play indicates the need to renew the bushes.
3 Remove the nut and spring washer from the end of the pivot bolt. Carefully tap the pivot bolt out of position, noting that an assistant should be to hand to help steady the machine as the bolt is withdrawn. Note that once this bolt is removed, the machine will be effectively in two parts, anchored only by the rear suspension units. Blocks should be used below the engine and between the side members and frame, to support the machine in a position where the bushes can be worked on. This will take a little experimentation but should not be unduly difficult to arrange. Make sure that the various cables are not placed under strain.
4 Prise out the dust seals from each end of the frame lug to expose the bushes. Drive the bushes out, using a long drift and hammer. There is a distance tube between the two bushes. This must be replaced when the new bushes are fitted. Tap one of the new bushes into position in the frame lug, then force some grease into the housing from the opposite side. Grease and fit the distance tube, then tap in the second bush. Ensure that the bushes are equidistant from the end of the tubular lug.
5 The grease seals/dust excluders can now be fitted to the ends of the lug. With assistance, the frame and side members can be realigned, and the pivot bolt refitted. Fit and tighten the securing nut.

10 Rear suspension units: removal and replacement

1 The rear swinging arm assembly is supported by the two suspension units. These units are each composed of a central rod, around which are arranged a large main spring and two smaller secondary springs. The large spring takes the main compression loading, assisted by one of the smaller springs, whilst the second smaller spring absorbs rebound shock. No form of hydraulic damping is employed.
2 The suspension units do not require any maintenance and can normally be expected to last the life of the machine. If renewal is required for any reason, it is recommended that they be replaced as a pair. Each unit is retained by a nut and bolt at its lower end and by a stud and nut at the top. Removal and replacement is a simple matter of releasing the mounting nuts and bolts, detaching the old unit and fitting a new one.

5.2a Stanchion is clamped in lower yoke

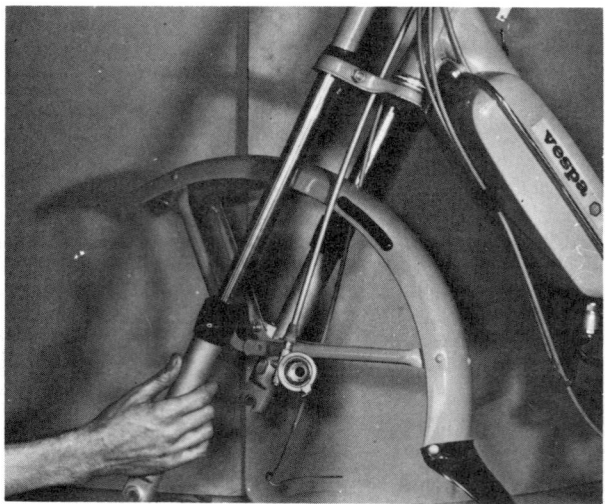

5.2b Remove fork leg, leaving shrouds in position

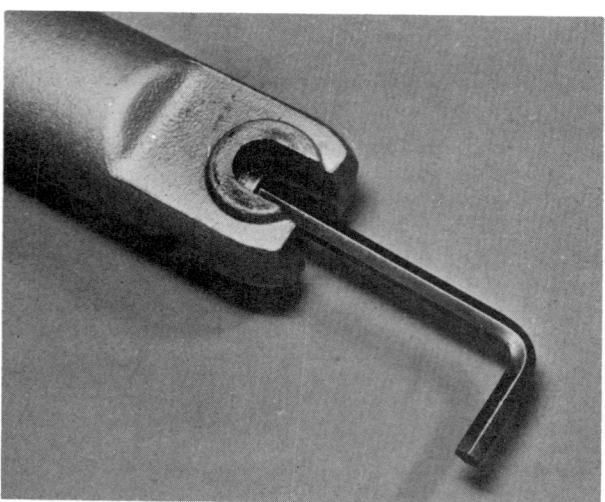

7.1 Release spring retaining block by unscrewing socket screw

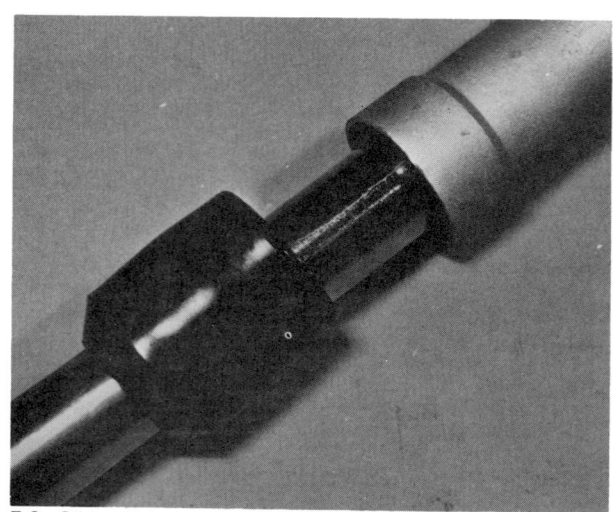

7.2a Slide off dust seal and separate fork assembly

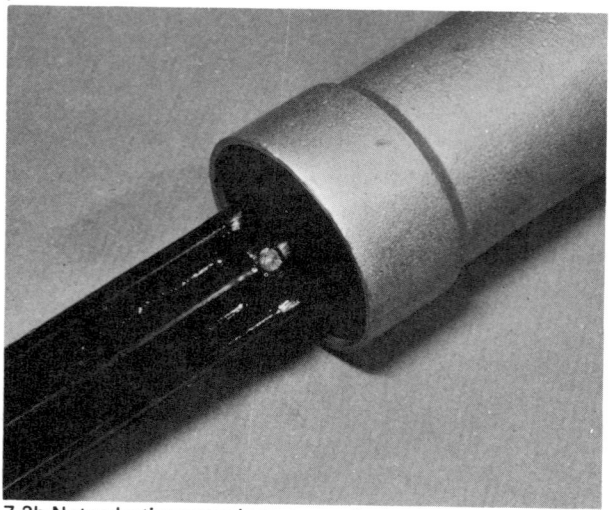

7.2b Note plastic spacer between springs

7.2c Retaining blocks thread into spring ends

Chapter 4 Frame and forks 55

7.3a Examine seal and renew, if worn

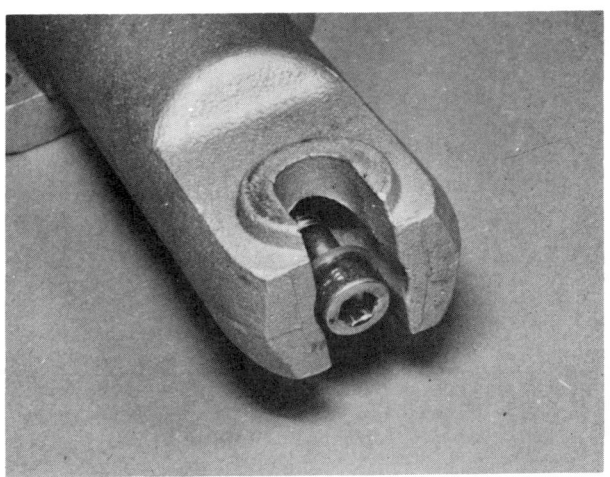

7.3b Socket screw through bottom of lower leg ...

7.3c ... secures alloy retaining block

7.3d Refill fork legs to level hole

11 Centre stand: examination

1 A tubular centre stand is fitted to support the machine when it is parked. It is supported on a long shaft retained by a circlip at each end. A return spring is fitted around the pivot shaft, to retract the stand when it is not in use.

2 Periodically, the pivot pin should be lubricated with grease. This is fairly important as its exposed position renders it susceptible to corrosion if left unattended.

3 Be especially careful to check the condition and correct location of the return spring. If this fails, the stand will fall onto the road in use, and may unseat the rider if it catches in a drain cover or similar obstacle.

12 Saddle: adjustment

1 The spring saddle is mounted on a stem, which is retained to the frame by a clamp arrangement. Provision is made for adjusting the height of the saddle, to cater for the individual requirements of owners. Adjustment can be carried out after the clamp bolt has been slackened. Position the saddle at the required height, then retighten the clamp bolt.

10.2 Suspension unit is retained by bolt and stud

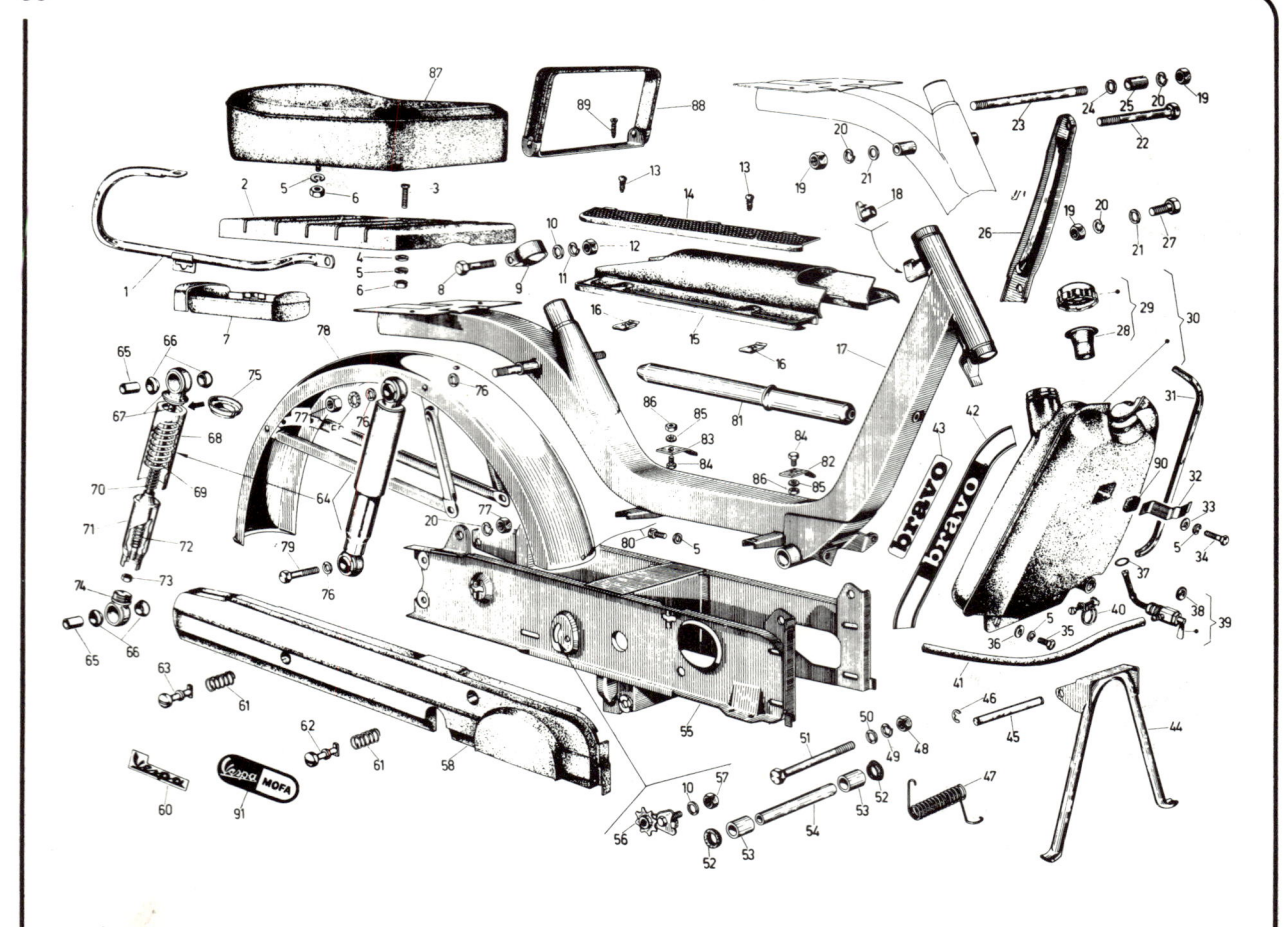

Fig.4.3 Bravo frame components

1 Pannier support rail	32 Cable clip	62 Short securing catch
2 Luggage rack	33 Plain washer	63 Long securing catch
3 Bolt – 2 off	34 Bolt	64 Rear suspension unit – 2 off
4 Plain washer – 2 off	35 Bolt – 2 off	65 Inner bush – 4 off
5 Spring washer – 6 off	36 Plain washer – 2 off	66 Rubber block – 8 off
6 Nut – 2 off	37 'O' ring	67 Centre rod – 2 off
7 Tool tray	38 Fuel tap seal	68 Upper cover – 2 off
8 Clamp bolt	39 Fuel tap assembly	69 Main spring – 2 off
9 Saddle clip	40 Securing clip	70 Inner spring – 2 off
10 Plain washer – 2 off	41 Petrol pipe	71 Lower cover – 2 off
11 Spring washer	42 Badge	72 Rebound spring – 2 off
12 Nut	43 Badge	73 Lower nut – 2 off
13 Self-tapping screw – 4 off	44 Stand	74 Bottom mounting lug
14 Footrest plate	45 Pivot pin	75 Retaining ring – 2 off
15 Top cover	46 Circlip – 2 off	76 Plain washer – 2 off
16 Retaining clip – 4 off	47 Stand spring	77 Nut – 4 off
17 Main frame member	48 Nut	78 Rear mudguard
18 Steering lock blanking plug	49 Spring washer – 3 off	79 Suspension unit mounting block – 2 off
19 Nut – 5 off	50 Plain washer	80 Bolt
20 Spring washer – 7 off	51 Pivot bolt	81 Tyre inflator
21 Plain washer – 3 off	52 Seal – 2 off	82 Bracket
22 Bolt	53 Bush – 2 off	83 Bracket
23 Stud	54 Spacer	84 Bolt – 2 off
24 Plain washer – 2 off	55 Side member assembly	85 Spring washer – 2 off
25 Spacer	56 Chain tensioner	86 Nut – 2 off
26 Rigid rear strut – 2 off	57 Nut	87 Optional passenger seat
27 Bolt – 2 off	58 Side cover (left and right-hand)	88 Grab handle
28 Seal	59 Shakeproof washer	89 Self-tapping screw – 2 off
29 Filler cap assembly	60 Badge	90 Rubber block
30 Fuel tank assembly	61 Spring – 4 off	91 Badge – 2 off
31 Fuel tank trim – 2 off		

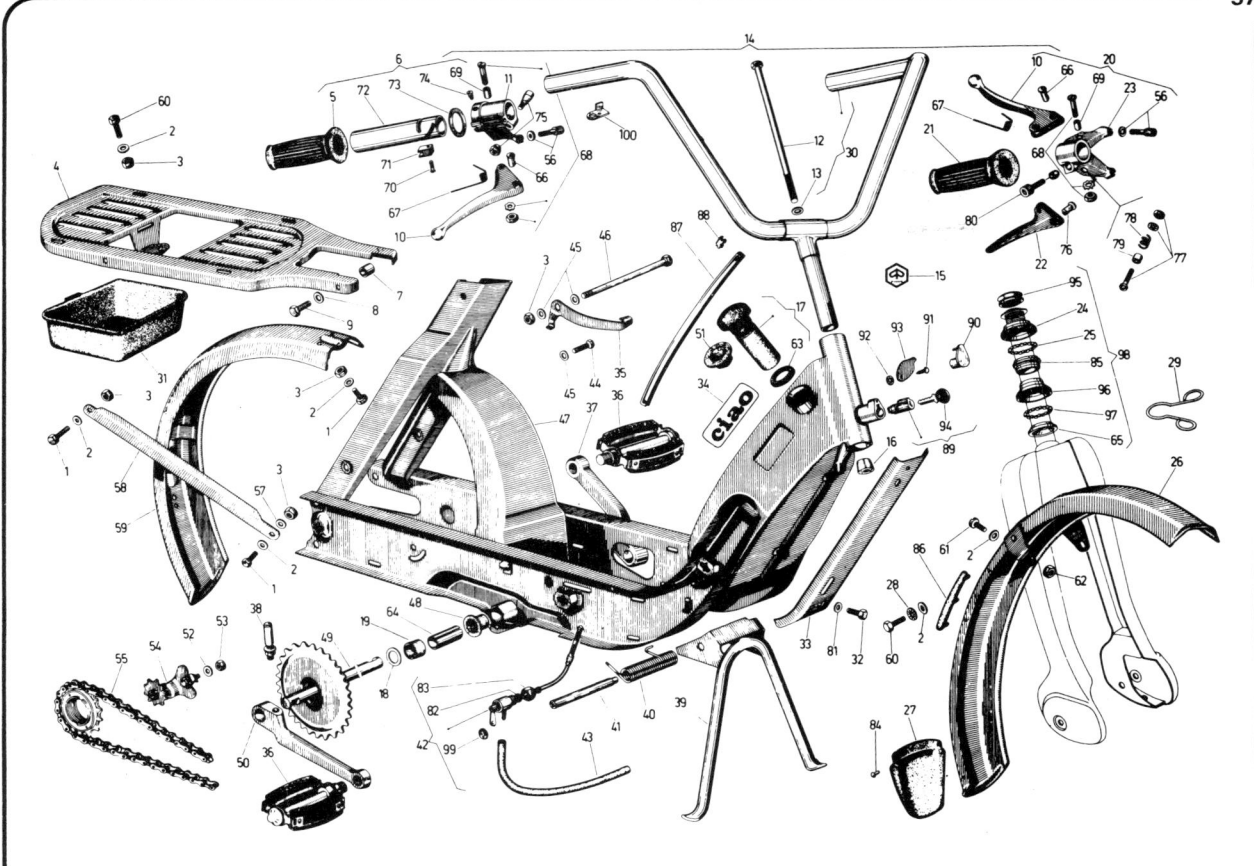

Fig. 4.4 Ciao frame components

1. Bolt – 6 off
2. Plain washer – 14 off
3. Nut – 10 off
4. Luggage rack
5. Right-hand handlebar grip
6. Throttle twist grip assembly
7. Spacer – 2 off
8. Plain washer – 4 off
9. Bolt – 2 off
10. Brake lever – 2 off
11. Lever pivot block
12. Steering column bolt
13. Plain washer
14. Handlebar assembly
15. Badge
16. Tapered wedge
17. Fuel filling cap
18. Thrust washer
19. Spacer
20. Lever assembly
21. Left-hand handlebar grip
22. Decompressor lever
23. Lever pivot block
24. Top cup
25. Caged ball race
26. Front mudguard
27. Mudflap
28. Shakeproof washer – 2 off
29. Cable guide
30. Handlebars
31. Tool tray
32. Screw – 2 off
33. Moulding
34. Badge
35. Engine adjustment lever
36. Pedal
37. Left-hand crank
38. Cotter pin – 2 off
39. Stand
40. Stand spring
41. Pivot shaft
42. Fuel tap assembly
43. Fuel pipe
44. Bolt
45. Plain washer – 5 off
46. Front engine bolt – 2 off
47. Main frame member
48. Bush – 2 off
49. Pedal shaft
50. Right-hand crank
51. Cup
52. Plain washer
53. Nut – 2 off
54. Chain tensioner
55. Pedal chain
56. Adjuster – 2 off
57. Plain washer – 2 off
58. Luggage rack stay – 2 off
59. Rear mudguard
60. Bolt – 4 off
61. Bolt – 2 off
62. Nut – 2 off
63. Seal
64. Spacer
65. Lower cone
66. Clevis
67. Spring – 2 off
68. Lever pivot assembly
69. Bush – 2 off
70. Screw
71. Sliding block
72. Twist grip sleeve
73. Washer
74. Grub screw
75. Clamp bolt
76. Clevis
77. Pivot assembly
78. Spring
79. Bush
80. Clamp screw
81. Plain washer – 2 off
82. Gasket
83. Gland nut
84. Rivet – 2 off
85. Upper cone
86. Rubbing strip
87. Trim
88. Clip
89. Steering lock
90. Blanking plug
91. Rivet
92. Washer
93. Cover
94. Steering lock key
95. Steering column nut
96. Lower cup
97. Caged ball race
98. Steering head bearing assembly
99. Fuel tap seal
100. Bracket

58

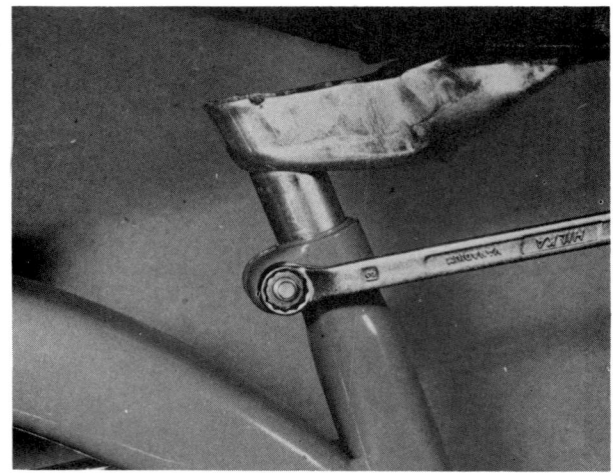

12.1 Saddle height can be adjusted

Fig. 4.5 Saddle suspension: Ciao Super Comfort model

1. Saddle spring
2. Saddle assembly
3. Luggage rack
4. Bush – 2 off
5. Spacer
6. Plain washer – 2 off
7. Shakeproof washer
8. Nut
9. Saddle clamp
10. Support beam
11. Frame
12. Saddle cover
13. Spring cover
14. Spacer – 2 off
15. Plain washer – 2 off
16. Bolt – 2 off
17. Pivot bolt
18. Spring centre bolt
19. Damping rubber
20. Saddle support spring
21. Spring seat
22. Nut
23. Badge denoting SC model

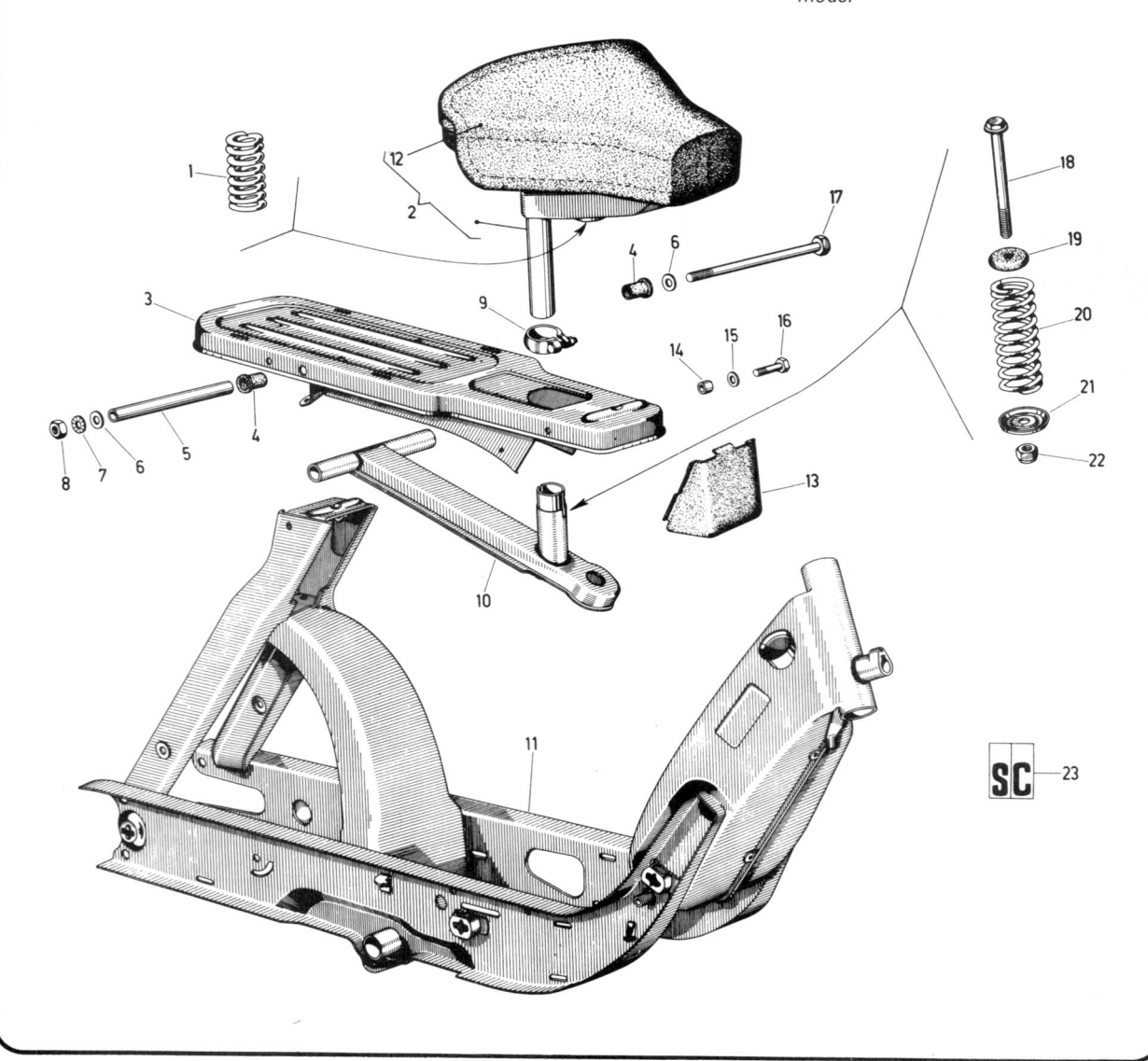

13 Speedometer head: removal and replacement

1 Release the screws which retain the upper half of the headlamp shell. Lift the shell half upwards so that the knurled ring securing the speedometer cable can be released. Slacken and remove the clamp which retains the speedometer to the headlamp shell; it can now be displaced from the shell half.

2 Apart from defects in either the drive or the drive cable, a speedometer or tachometer that malfunctions is difficult to repair. Fit a new one, or alternatively entrust the repair to a competent instrument repair specialist.

14 Speedometer drive cable: examination and lubrication

1 It is advisable to detach the cable from time to time in order to check whether it is lubricated adequately, and whether the outer covering is compressed or damaged at any point along its run. Jerky or sluggish movements can often be attributed to a cable fault.

2 For greasing, withdraw the inner cable. After wiping off the old grease, clean with a petrol-soaked rag and examine the cable for broken strands or other damage.

3 Regrease the cable with high melting point grease, taking care not to grease the last six inches at the point where the cable enters the instrument head. If this precaution is not observed, grease will work into the head and immobilise the movement.

4 If the speedometer ceases to function, suspect a broken cable. Inspection will show whether the inner cable has broken; if so, the inner cable alone can be renewed and resinserted in the outer casing after greasing. Never fit a new inner cable alone if the outer covering is damaged or compressed at any point.

15 Cleaning the machine

1 After removing all surface dirt with a rag or sponge washed frequently in clean water, the machine should be allowed to dry thoroughly. Application of car polish or wax to the cycle parts will give a good finish, particularly if the machine has been neglected for a long period.

2 The plated parts of the machine should require only a wipe with a damp rag. If the plated parts are badly corroded, as may occur during the winter when the roads are salted, it is preferable to use one of the proprietary chrome cleaners. These often have an oily base, which will help to prevent the corrosion from re-occurring.

3 If the engine parts are particularly oily, use a cleaning compound such as 'Gunk' or 'Jizer'. Apply the compound whilst the parts are dry and work it in with a brush so that it has the opportunity to penetrate the film of grease and oil. Finish off by washing down liberally with plenty of water, taking care that it does not enter the carburettor or the electrics. If desired, the now clean aluminium alloy parts can be enhanced further by using a special polish such as Solvol Autosol, which will fully restore their brilliance.

4 Whenever possible, the machine should be wiped down after it has been used in the wet, so that it is not garaged under damp conditions which will promote rusting. Remember there is little chance of water entering the control cables and causing stiffness of operation if they are lubricated regularly as recommended in the Routine Maintenance Section.

13.1a Speedometer is mounted in headlamp shell ...

13.1b ... and is retained by nut and clamp

Chapter 4 Frame and forks

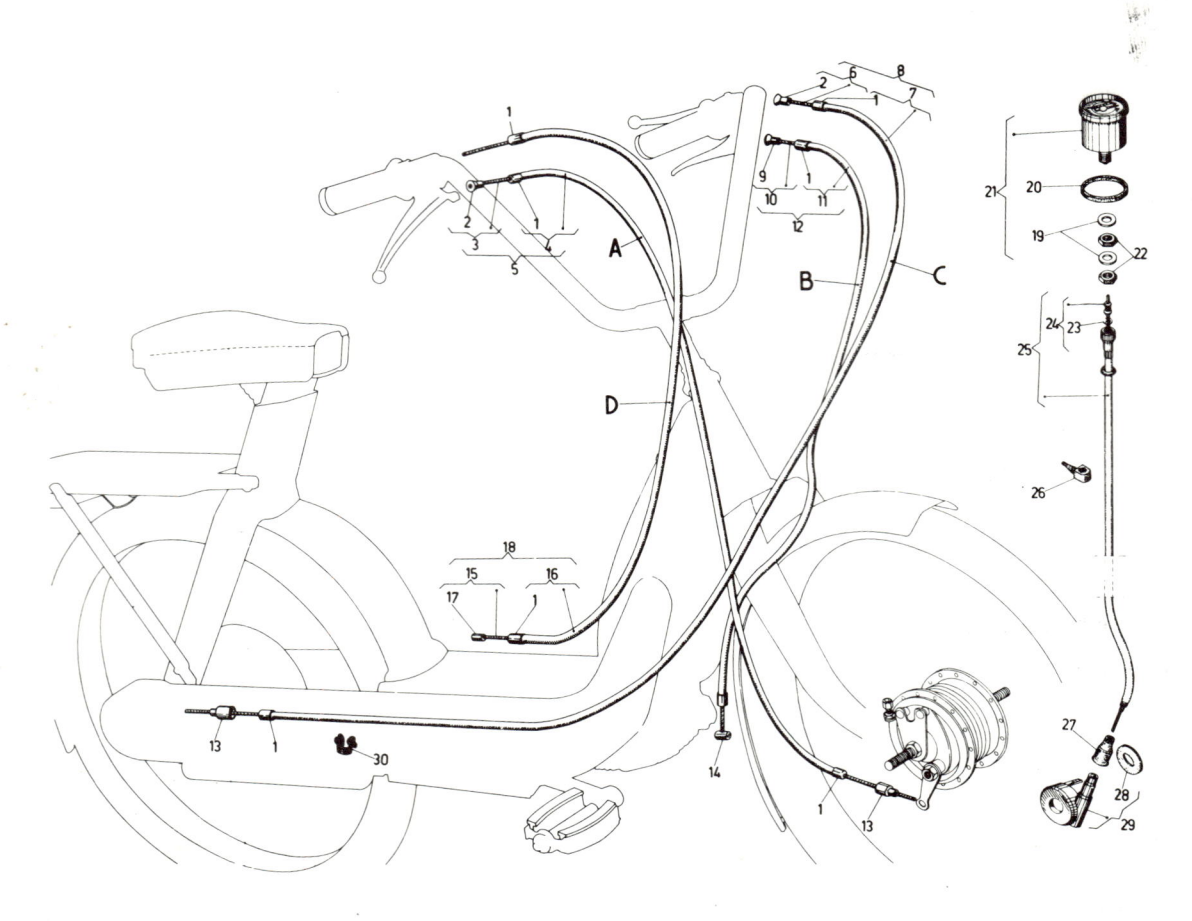

Fig. 4.6 Control cables and speedometer

- A Front brake cable complete
- B Decompressor cable complete
- C Rear brake cable complete
- D Throttle cable complete

1. Ferrule – 8 off
2. Nipple – 8 off
3. Front brake cable – inner
4. Front brake cable – outer
5. Front brake cable – complete
6. Rear brake cable – inner
7. Rear brake cable – outer
8. Rear brake cable – complete
9. Nipple
10. Decompressor cable – inner
11. Decompressor cable – outer
12. Decompressor cable – complete
13. Ferrule
14. Barrel nipple
15. Throttle cable – inner
16. Throttle cable – outer
17. Nipple
18. Throttle cable – complete
19. Plain washer – 2 off
20. Rubber ring
21. Speedometer head complete
22. Nut – 2 off
23. Washer
24. Speedometer cable – inner
25. Speedometer drive cable – complete
26. Cable guide
27. Dust cap
28. Washer
29. Speedometer drive assembly
30. Cable clip

Symptom	Cause	Remedy
Machine veers either to the left or the right with hands off handlebars	Bent frame Twisted forks	Check, and renew. Check, and renew.
Machine rolls at low speed	Overtight steering head bearings	Slacken until adjustment is correct.
Machine judders when front brake is applied	Slack steering head bearings	Tighten until adjustment is correct.
Fork action stiff (hydraulic fork models only)	Fork legs out of alignment (twisted in yokes)	Slacken yoke clamps, and fork top bolts. Pump fork several times then retighten from bottom upwards.
Machine wanders. Steering imprecise Rear wheel tends to hop	Worn swinging arm pivot (Bravo models only)	Dismantle and renew bushes and pivot shaft.

Chapter 5 Wheels, brakes and tyres

Contents

General description ... 1	Rear wheel: examination and renovation ... 8
Front wheel: examination and renovation ... 2	Rear wheel: removal ... 9
Front wheel: removal, all models ... 3	Rear wheel bearings: examination and renovation ... 10
Front brake: examination and renovation ... 4	Adjusting the front and rear brakes ... 11
Front wheel bearings: examination and renovation ... 5	Rear brake shoe: examination ... 12
Speedometer drive gearbox: examination ... 6	Tyres: removal and replacement ... 13
Front wheel: replacement ... 7	Fault diagnosis: wheels, brakes and tyres ... 14

Specifications

	Bravo		Ciao
Tyre size: front	2.50 x 16 in	or	2.00 x 17 in
Tyre size: rear	2.50 x 16 in	or	2.00 x 17 in
Tyre pressure: front	17 psi (1.2 kg/cm^2)		20 psi (1.4 kg/cm^2)
Tyre pressure: rear	28 psi (2.0 kg/cm^2)		35.5 psi (2.5 kg/cm^2)
Wheel type	Steel rim, spoked		

Brakes

Type	front	Internal expanding, single leading shoe
	rear	Internal expanding, single shoe

1 General description

All models are equipped with steel rims, laced to pressed steel hubs by way of spokes. The front hub incorporates a drum brake of the single leading shoe (SLS) type operated by Bowden cable from a handlebar lever. The rear hub contains the hub unit casing which fits into a recess in the drum. This unit is fitted with a single rear brake shoe. This is also operated from a handlebar lever, by a Bowden cable.

2 Front wheel: examination and renovation

1 Place the machine on the centre stand so that the front wheel is raised clear of the ground. Spin the wheel and check the rim alignment. Small irregularities can be corrected by tightening the spokes in the affected areas, although a certain amount of practice is necessary to prevent over-correction. Any flats in the wheel rim should be evident at the same time. These are more difficult to remove and in most cases it will be necessary to have the wheel rebuilt on a new rim. Apart from the effects on stability, a flat will expose the tyre bead and walls to greater risk of damage.

2 Check for loose or broken spokes. Tapping the spokes is the best guide to tension. A loose spoke will produce a quite different sound and should be tightened by turning the nipple in an anti-clockwise direction. Always re-check for run-out by spinning the wheel again. If the spokes have to be tightened an excessive amount, it is advisable to remove the tyre and tube by the procedure detailed in Section 13 of this Chapter; this is so that the protruding ends of the spokes can be ground off, to prevent them from chafing the inner tube and causing punctures.

3 Front wheel: removal, all models

1 With the machine supported on the centre stand, block the stand and crankcase to raise the wheel clear of the ground.
2 Detach the speedometer drive cable at the wheel by unscrewing the knurled gland nut which retains it. Disconnect the front brake cable at the end of the actuating arm, and position both cables out of the way of the wheel.
3 Slacken the wheel spindle nuts at each side of the wheel and lower the wheel clear of the forks. It can now be removed for further attention.

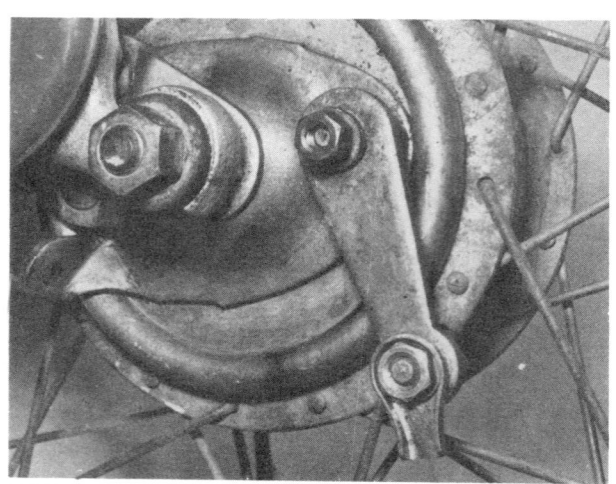

3.2 Release front brake cable and wheel nuts

4 Front brake: examination and renovation

1 To remove the brake assembly, unscrew the nut on the spindle and lift it out from the brake drum.
2 Examine the brake linings. If they are wearing thin or unevenly, the brake shoes should be renewed. The linings are bonded on and cannot be replaced as a separate item.
3 To remove the brake shoes from the brake plate assembly, arrange the operating lever so that brakes are in the 'full on' position and then pull the shoes apart whilst lifting them upward in the form of a 'V'. When they are clear of the brake plate, the return springs can be removed and the shoes separated.
4 Before replacing the brake shoes, check that the brake operating cam is working smoothly and is not binding in its pivot. The cam can be removed by withdrawing the retaining nut on the operating arm and pulling the arm off the shaft. Before removing the arm, it is advisable to mark its position in relation to the shaft, so that it can be relocated correctly. The shaft should be greased prior to reassembly and also a light smear of grease placed on the faces of the operating cam.
5 Check the inner surface of the brake drum on which the brake shoes bear. The surface should be smooth and free from score marks or indentations, otherwise reduced braking efficiency will be inevitable. Remove all traces of brake lining dust and wipe with a clean rag soaked in petrol to remove any traces of grease or oil.
6 To reassemble the brake shoes on the brake plate, fit the return springs first and then force the shoes apart, holding them in a 'V' formation. If they are now located with the brake operating cam and pivot they can usually be snapped into position by pressing downward. Never use excessive force, otherwise there is a risk of distorting the shoes permanently.

5 Front wheel bearings: examination and renovation

1 Access to the wheel bearings is gained when the brake assembly has been removed. There is a seal on both sides to prevent the grease from leaking out, especially into the brake drum. This seal should be prised off.
2 The wheel bearings are of the cup and cone type, each wheel containing 20 loose ball bearings. Adjustment is effected from the brake drum side of the wheel by means of a locknut in front of the threaded cone. The locknut can be slackened by unscrewing (right-hand thread), whilst the cone is held steady by means of a thin spanner across the two flats on its outer surface. When the locknut has been withdrawn, the cone can be unscrewed from the spindle, exposing the ball bearings and the cup in which they seat. Remove the ball bearings and place them in a safe place for further attention.
Turn the wheel over and withdraw the spindle complete with its cone. The right-hand cup and its contents of ball bearings is now also exposed.
3 Remove all the old grease from the bearings and wash the ball bearings in petrol, keeping the right and left hand assemblies apart. Check both sets of cups and cones for wear or discolouration and examine also the loose ball bearings. Ball bearings are cheap and if any show signs of defect, the entire assembly should be renewed without question. The cups and cones should have a polished appearance in the area of the bearing tracks. New replacement should be fitted if any surface defects are evident. The cups are a light drive fit into the hub and can be driven out with a suitable size of drift. When driving the new replacements into position, pressure should be applied only to the outer edge of the cups, to obviate the risk of distortion or damage.
4 Pack the cups with high melting point grease and reposition the ball bearings. Note that each race has a gap which gives the impression that one ball bearing is missing. This gap is deliberate, to give the bearings room to roll rather than skid against each other. Fit the spindle from the right-hand side of the wheel, after coating the cone with grease. Invert the wheel and fit the left-hand cone and locknut, after greasing the former. Adjust the cone so that the spindle will revolve freely with just the slightest amount of play at the wheel rim. Tighten the locknut and recheck the adjustment. If there is no free play, the bearings will absorb a surprising amount of power and overheat. Too much play will cause imprecise handling when the machine is back on the road.

6 Speedometer drive gearbox: examination

1 Before replacing the speedometer drive gearbox, it should be inspected to make sure it is well greased and that no part of the drive mechanism is faulty.
2 The speedometer drive gearbox rarely gives trouble. In the event of damage, it is not possible to effect a satisfactory repair; renewal is on the only solution.

7 Front wheel: replacement

1 To replace the front wheel, reverse the removal procedure and ensure that the peg on the forks locates in the slot in the brake plate. The need for this cannot be overstressed. If the brake plate is not anchored in this manner, the brake will lock on immediately it is applied, which may well result in a serious accident.
2 Reconnect the front brake and check that the brake functions correctly, especially if the adjustment has been altered or the brake operating arm has been removed and replaced during the dismantling operation.
3 Reconnect the speedometer drive.

8 Rear wheel: examination and renovation

1 Place the machine on the centre stand, so that the rear wheel is clear of the ground. Check the wheel for rim alignment, damage to the rim or loose or broken spokes, by following the procedure adopted for the front wheel in Section 2 of this Chapter.

Tyre removal: Deflate inner tube and insert lever in close proximity to tyre valve

Use two levers to work bead over the edge of rim

When first bead is clear, remove tyre as shown

Tyre fitting: Inflate inner tube and insert in tyre

Lay tyre on rim and feed valve through hole in rim

Work first bead over rim, using lever in final section

Use similar technique for second bead, finish at tyre valve position

Push valve and tube up into tyre when fitting final section, to avoid trapping

4.1 Remove brake assembly for examination

4.2 Bonded shoes should be renewed, if worn

5.3 Balls are retained by steel ring

5.4a Fit wheel spindle and cones ...

5.4b ... then adjust for free play

7.1a Hole in drum is for speedometer drive

Chapter 5 Wheels, brakes and tyres

7.1b Ensure that torque arm is correctly engaged

7.1c Refit front brake cable in pressed steel cleat

9 Rear wheel: removal

1 Rear wheel removal on both Ciao and Bravo models is a somewhat complicated function due to the belt transmission and the incorporation of the final reduction gears, or hub unit, in the rear wheel itself. It should be noted that it is possible to repair a puncture without actually removing the wheel, but any other work, like tyre changing or rear brake maintenance, will necessitate rear wheel removal.

2 Place the machine on its centre stand, and remove both of the plastic side covers. Slacken off the pedal chain tensioner by releasing the nut on the inside of the right-hand side member. The tensioner can now be swung clear of the chain, allowing it to be lifted off the freewheel unit. It is not normally necessary to have to separate the chain at the joining link in order that it can be removed.

3 On fixed ratio models, release the three engine mounting bolts, and move the engine unit fully back to allow the drive belt to be disengaged from the rear pulley. On variable ratio machines, slacken the two front mounting bolts and remove the short rear bolt entirely, to allow the unit to be moved back. The belt can now be removed from the front pulley. Release the rear brake cable from its cleat on the end of the actuating arm.

4 The rear wheel is secured by two bolts at each side which pass through the end of the side members. Remove the bolts and pull the wheel back as far as the mudguard will allow it to go. It will now be necessary to turn the hub unit clockwise (viewed from the left-hand side) as the wheel is lowered, so that the input shaft clears the cutout in the end of the side member. The wheel assembly can then be removed to await further dismantling.

10 Rear wheel bearings: examination and renovation

1 The rear wheel is supported on a roller bearing carried in the end of the hub unit, and a ball bearing in a separate bearing housing which incorporates the mounting lugs for rear wheel attachment. In order to gain access to the roller bearing, it will be necessary to remove and dismantle the hub unit as described in Chapter 1, Part 2, Section 4.

2 Wash the bearing out thoroughly with clean petrol, and allow it to dry or blow it dry with compressed air. The bearing can be checked for wear by spinning it, which should cause it to turn smoothly with no signs of rough spots or play. In practice, it is unlikely that this bearing will exhibit very much wear as it is completely enclosed by the hub unit and runs in gear oil at a relatively low speed.

3 Many owners may prefer to entrust the renewal operation to a Vespa Service Agent. The operation is, however, fairly simple providing reasonable care is taken; Support the hub casing, open end uppermost, on wooden blocks placed near to the bearing boss. The bearing can be tapped out of the casing, using a large socket. Should it prove to be a tight fit, the casing can be heated in an oven to 100°c, thus expanding the alloy and releasing the grip on the bearing. The oil seal, which must be renewed as a matter of course when it has been disturbed, will be pushed out by the bearing.

4 The new bearing can be fitted in the reverse order of that described above, ensuring that it seats squarely in the housing bore. A new seal should be tapped very carefully into position, and finally, the hub unit reassembled.

5 The ball bearing is much easier to renew, as it can be dealt with without removing the wheel completely, after releasing the pedal chain, freewheel and boss, and the two retaining bolts on the right-hand side of the wheel.

6 The bearing can be displaced from the housing after the retaining plate has been removed. It is retained by two rivets, which can be punched or drilled out. The bearing should be checked in the same way as that described for the roller bearing, after it has been tapped out of the housing.

7 Reassembly is a direct reversal of the removal process, ensuring that the retaining plate is refitted in the correct position. New rivets must be fitted, preferably using a concave-ended punch to retain them.

11 Adjusting the front and rear brakes

1 The front and rear brakes are operated by the right and left-hand brake levers respectively, each of which is provided with a threaded adjuster in order that brake shoe wear can be compensated for by effectively reducing the length of the operating cable. The setting of the brake levers is largely a matter of discretion on the part of the individual owner, but as a general guide, it can be considered that each brake should commence operation as soon as the lever is applied. When setting the brakes, check that the shoes free off completely as any slight binding will have a marked effect on the performance of the machine.

2 Should the range of adjustment prove insufficient, it is possible to shorten the cable at the point at which it is attached to the brake actuating arm. Release the screw and nut which clamps the inner cable to the small pressed steel cleat. The cable can now be disengaged and repositioned in the cleat, and the securing nut tightened. Final fine adjustment should be carried out in the normal way.

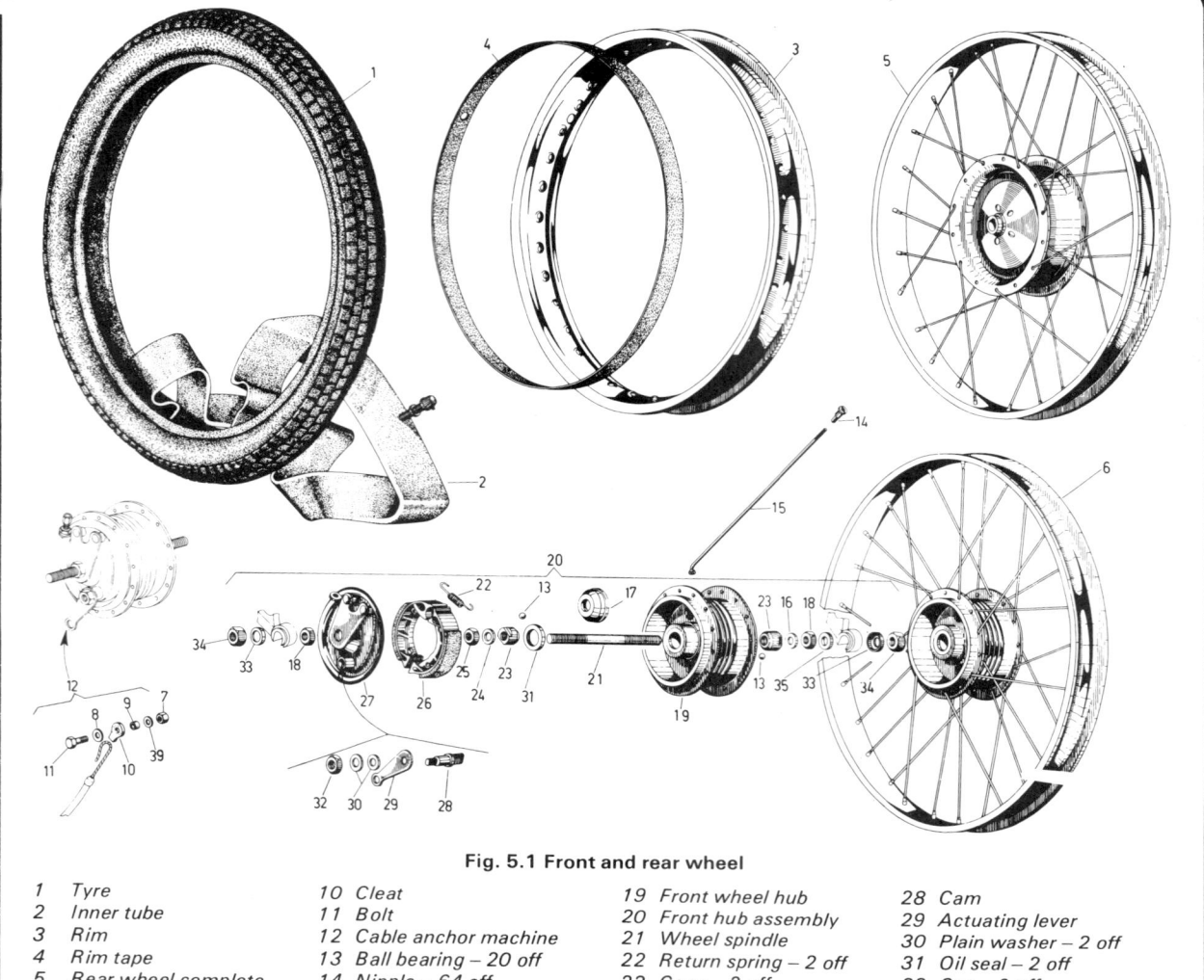

Fig. 5.1 Front and rear wheel

1 Tyre
2 Inner tube
3 Rim
4 Rim tape
5 Rear wheel complete
6 Front wheel complete
7 Nut
8 Plain washer – 2 off
9 Spacer
10 Cleat
11 Bolt
12 Cable anchor machine
13 Ball bearing – 20 off
14 Nipple – 64 off
15 Spoke – 64 off
16 Spacer – 2 off
17 Cup – 2 off
18 Lock nut – 2 off
19 Front wheel hub
20 Front hub assembly
21 Wheel spindle
22 Return spring – 2 off
23 Cone – 2 off
24 Spacer
25 Lock nut
26 Brake shoes – 2 off
27 Brake plate
28 Cam
29 Actuating lever
30 Plain washer – 2 off
31 Oil seal – 2 off
33 Cup – 2 off
34 Wheel spindle nut – 2 off
35 Spacer

9.3 Release rear brake cable at actuating arm

9.4a Wheel is retained by two bolts at each side

Chapter 5 Wheels, brakes and tyres

9.4b Twist wheel as it is lowered, to clear spindle

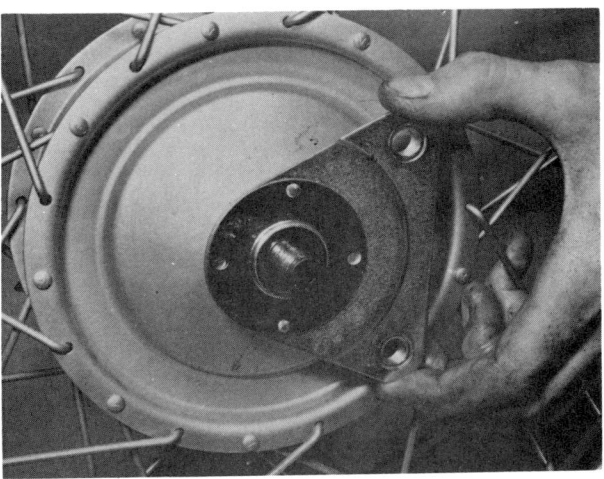

10.5 Ball race is enclosed in plate

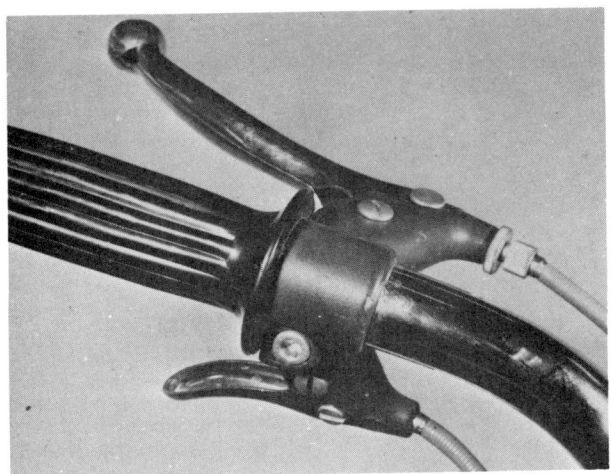

11.1 Adjust cable free play by way of adjusters

12 Rear brake shoe: examination

1 The single rear brake shoe is mounted on the back of the hub casing and is operated by a simple actuating arm to which the cable is attached. One end of the shoe is retained to its pivot pin by a small spring clip, the other end being held by a return spring. The shoe is of the bonded type, and should be renewed complete when the lining material shows signs of extensive wear.

12.1a Rear brake has single brake shoe

12.1b Lever is retained by circlip

Chapter 5 Wheels, brakes and tyres

13 Tyres: removal and replacement

1 At some time or other the need will arise to remove and replace the tyres, either as the result of a puncture or because a replacement is required to offset wear. To the inexperienced, tyre changing represents a formidable task yet if a few simple rules are observed and the technique learned, the whole operation is surprisingly simple.

2 To remove the tyre from either wheel, first detach the wheel from the machine by following the procedure in Section 3 or Section 9 of this Chapter, depending on whether the front or the rear wheel is involved. It should be noted that it is possible to repair a rear wheel puncture in situ, thus avoiding a considerable amount of work involved in removing the wheel. Further attention, or tyre replacement will, however, necessitate the removal of the wheel. Deflate the tyre by removing the valve insert and when it is fully deflated, push the bead of the tyre away from the wheel rim on both sides so that the bead enters the centre well of the rim. Remove the locking cap and push the tyre valve into the tyre itself.

3 Insert a tyre lever close to the valve and lever the edge of the tyre over the outside of the wheel rim. Very little force should be necessary; if resistance is encountered it is probably due to the fact that the tyre beads have not entered the well of the wheel rim all the way round the tyre.

4 Once the tyre has been edged over the wheel rim, it is easy to work around the wheel rim so that the tyre is completely free on one side. At this stage, the inner tube can be removed.

5 Working from the other side of the wheel, ease the other edge of the tyre over the outside of the wheel rim that is furthest away. Continue to work around the rim until the tyre is free completely from the rim.

6 If a puncture has necessitated the removal, reinflate the inner tube and immerse it in a bowl of water to trace the source of the leak. Mark its position and deflate the tube. Dry the tube and clean the area around the puncture with a petrol soaked rag. When the surface has dried, apply the rubber solution and allow this to dry before removing the backing from the patch and applying the patch to the surface.

7 It is best to use a patch of the self vulcanising type, which will form a very permanent repair. Note that it may be necessary to remove a protective covering from the top surface of the patch, after it has sealed in position. Inner tubes made from synthetic rubber may require a special type of patch and adhesive, if a satisfactory bond is to be achieved.

8 Before replacing a tyre, check the inside to make sure the agent that caused the puncture is not trapped. Check also the outside of the tyre, particularly the tread area, to make sure nothing is trapped that may cause a further puncture.

9 If the inner tube has been patched on a number of past occasions, or if there is a tear or large hole, it is preferable to discard it and fit a replacement. Sudden deflation may cause an accident, particularly if it occurs with the front wheel.

10 To replace the tyre, inflate the inner tube sufficiently for it to assume a circular shape but only just. Then put it into the tyre so that it is enclosed completely. Lay the tyre on the wheel at an angle and insert the valve through the rim tape and the hole in the wheel rim. Attach the locking cap on the first few threads, sufficient to hold the valve captive in its correct location.

11 Starting at the point furthest from the valve, push the tyre bead over the edge of the wheel rim until it is located in the central well. Continue to work around the tyre in this fashion until the whole of one side of the tyre is on the rim. It may be necessary to use a tyre lever during the final stages.

12 Make sure there is no pull on the tyre valve and again commencing with the area furthest from the valve, ease the other bead of the tyre over the edge of the rim. Finish with the area close to the valve, pushing the valve up into the tyre until the locking cap touches the rim. This will ensure the inner tube is not trapped when the last section of the bead is edged over the rim with a tyre lever.

13 Check that the inner tube is not trapped at any point. Reinflate the inner tube and check that the tyre is seating correctly around the wheel rim. There should be a thin rib moulded around the wall of the tyre on both sides, which should be equidistant from the wheel rim at all points. If the tyre is unevenly located on the rim, try bouncing the wheel when the tyre is at the recommended pressure. It is probable that one of the beads has not pulled clear of the centre well.

14 Always run the tyres at the recommended pressures and never under or over-inflate. The correct pressures for solo use are given in the Specifications Section of this Chapter.

15 Tyre replacement is aided by dusting the side walls, particularly in the vicinity of the beads, with a liberal coating of French chalk. Washing up liquid can also be used to good effect, but this has the disadvantage of causing the inner surfaces of the wheel rim to rust.

16 Never replace the inner tube and tyre without the rim tape in position. If this precaution is overlooked there is a good chance of the ends of the spoke nipples chafing the inner tube and causing a crop of punctures.

17 Never fit a tyre that has a damaged tread or side walls. Apart from the legal aspects, there is a very great risk of a blow-out, which can have serious consequences on any two-wheel vehicles.

18 Tyre valves rarely give trouble, but it is always advisable to check whether the valve itself is leaking before removing the tyre. Do not forget to fit the dust cap, which forms an effective second seal.

14 Fault diagnosis – wheels, brakes and tyres

Symptom	Cause	Remedy
Handlebars oscillate at low speeds	Buckled front wheel Incorrectly fitted front tyre	Remove wheel for specialist attention. Check whether line around bead is equidistant from rim.
Brakes ineffective	Contaminated or glazed linings	Remove and renew or remove glaze as necessary.
Brakes feel spongy	Cable badly routed Stretched brake operating cables, weak pull-off springs	Re-route cable(s) avoiding sharp bends. Renew cables and/or springs, after inspection.
Tyres wear more rapidly in middle of tread	Over inflation	Check pressures and run at recommended settings.
Tyres wear rapidly at outer edges of tread	Under inflation	Ditto.

Chapter 6 Electrical system

Contents

General description	1	Tail lamp: replacing bulb	5
Checking the electrical system: general	2	Horn: location	6
Headlamp: replacing bulbs and adjusting the beam height	3	Wiring: layout and examination	7
Horn and light switches: location and renovation	4	Fault diagnosis: Electrical system	8

Specifications

Generator
Type ... 6 volt flywheel magneto/AC generator

Bulbs
Headlamp ... 6V 15W (6V 20W – later models)
Tail lamp ... 6V 3W festoon
Pilot lamp ... 6V 15W festoon

Horn
Type ... 6V ac

1 General description

The electrical system on all Vespa mopeds is powered by a flywheel magneto/generator unit which produces 6 volt alternating current (ac). The ac current is fed directly to the electrical system, where it is used to provide lighting and to operate the ac electric horn. The system is so designed that the full complement of bulbs fitted on assembly is capable of handling the output of the generator at all normal engine speeds.

2 Checking the electrical system: general

1 It is unlikely that anything serious will go wrong with the electrical system during the normal life of the machine, apart from the more common failures such as bulbs or corroded switch contacts. Few owners will have the facilities to diagnose and repair the generator coils, should these fail, and if elimination of the rest of the system indicates a failure in the generator itself, it is recommended that it be entrusted to a Vespa Service Agent or an Auto-Electrician who will have the equipment and expertise necessary to effect an economical repair.

1.1 Flywheel magneto components:
A Contact breaker assembly   C Lighting coil
B Condenser   D Ignition low tension coil

Chapter 6 Electrical system

3 Headlamp: replacing bulbs and adjusting the beam height

1   The top half of the headlamp unit can be released after unscrewing the two screws which pass up from the bottom of the assembly, (see photographs). The headlamp bulb is retained by the base contact, and can be withdrawn after this has been swivelled to one side.
2   A festoon-type pilot bulb, when fitted to provide speedometer illumination, is clipped between two contacts. When fitting a new bulb, ensure that it is gripped firmly between them. It should be noted that the most common cause of blown bulbs on any direct lighting system is bad bulb connections, which allow spasmodic overloading of the system. Similarly, if one bulb blows, it is likely that the overload which will result will soon damage the remainder, and for this reason it is wise to carry a complete set of spare bulbs for emergencies.
3   Beam alignmernt is achieved by pivoting the whole lamp between the mounting brackets or by slackening two screws in the base, and moving the lens/reflector unit. Examination will soon show which type of unit is fitted. The beam should be set to give a good spread of light about 15 - 20 feet from the machine. There is nothing to be gained by setting the beam too high as the low wattage of the bulb will mean that most of the available light is scattered and will do no good at all.
4   Always fit replacement bulbs of the same wattage of those fitted. Note that a higher wattage bulb in this type of system will usually produce **less** light than the standard one, whilst a lower wattage bulb will overload, fusing the filament.

4 Horn and light switches: location and renovation

1   Various types of switch have been incorporated on Ciao and Bravo models. Ciao models make use of a handlebar-mounted switch which incorporates the light switch and horn push. Bravo models have a sliding switch mounted on the top of the headlamp unit, and a separate horn push, which forms part of the twistgrip unit.
2   The main cause of switch failure is dirty or corroded contacts which, if allowed to develop, may ultimately cause intermittent isolation of one or more of the electrical components. This can lead in turn to problems with bulbs fusing at frequent intervals. The best form of maintenance is to keep the contacts covered with petroleum jelly, to exclude moisture. (**Do not** use grease on electrical contacts). The switch can be cleaned out using a proprietary switch cleaning aerosol spray, if necessary.
3   It is not normally possible to repair a switch if mechanical breakage occurs, and it is recommended that a replacement unit is fitted in this eventuality.

3.1a Note securing screws (A) and adjusting screws (B)

3.1b Earlier type has securing screws only (arrowed)

3.1c Swing contact aside to release headlamp bulb

3.2 Pilot festoon bulb is clipped between contacts

Chapter 6 Electrical system

4.1a Keep contacts clean. Note label showing colour coding

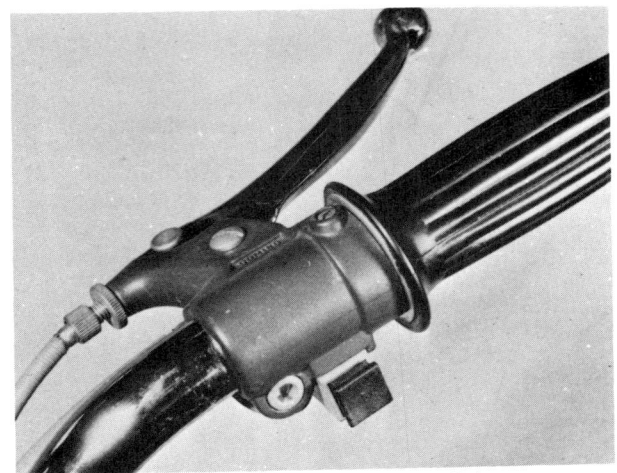

4.1b Late models have horn push only on handlebars

5 Tail lamp: replacing bulb

1   The moulded plastic cover of the rear lamp is retained by two screws. When these screws are removed, the cover can be removed and the bulb exposed.
2   The tail lamp bulb is of the festoon type, and is retained between two spring contacts.
3   Consistent problems with bulbs blowing may be traced to a faulty earth or feed connection. Keep the main connection bright by cleaning with a fine emery strip.

6 Horn: location

The horn is usually incorporated in the headlamp assembly, and may be removed after the top of the headlamp unit has been detached. In the event that the horn malfunctions, it should be renewed. There is no satisfactory method of repairing the unit.

5.2 Rear lamp uses small festoon bulb

6.1 Horn is incorporated in headlamp assembly

7 Wiring: layout and examination

1   The wiring harness is colour-coded and will correspond with the accompanying wiring diagram.
Visual inspection will show whether there are any breaks or frayed outer coverings which will give rise to short circuits. Another source of trouble may be the snap connectors and sockets, where the connector has not been pushed fully home in the outer housing.
Intermittent short circuits can often be traced to a chafed wire that passes through or is close to a metal component such as a frame member. Avoid tight bends in the lead or situations where a lead can become trapped between casings.

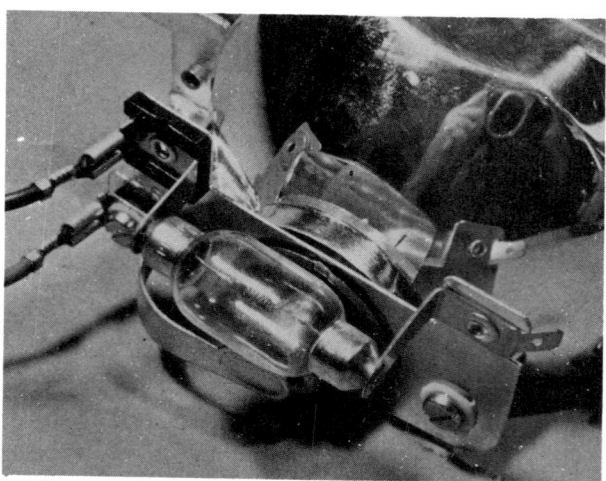

7.1 Colour coded cables are attached by Lucar terminals

8 Fault diagnosis – electrical system

Symptom	Cause	Remedy
Dim lights, horn inoperative	Poor connections, dirty switch contacts	Check and clean.
	Bulb wattage too high	Check and fit correct bulb.
	Flywheel demagnetized	Remagnetize flywheel
Constantly blowing bulbs	Vibration, intermittent feed or earth connection	Check and rectify as necessary.
	Incorect bulbs fitted	Fit bulbs of correct value

Wiring diagram

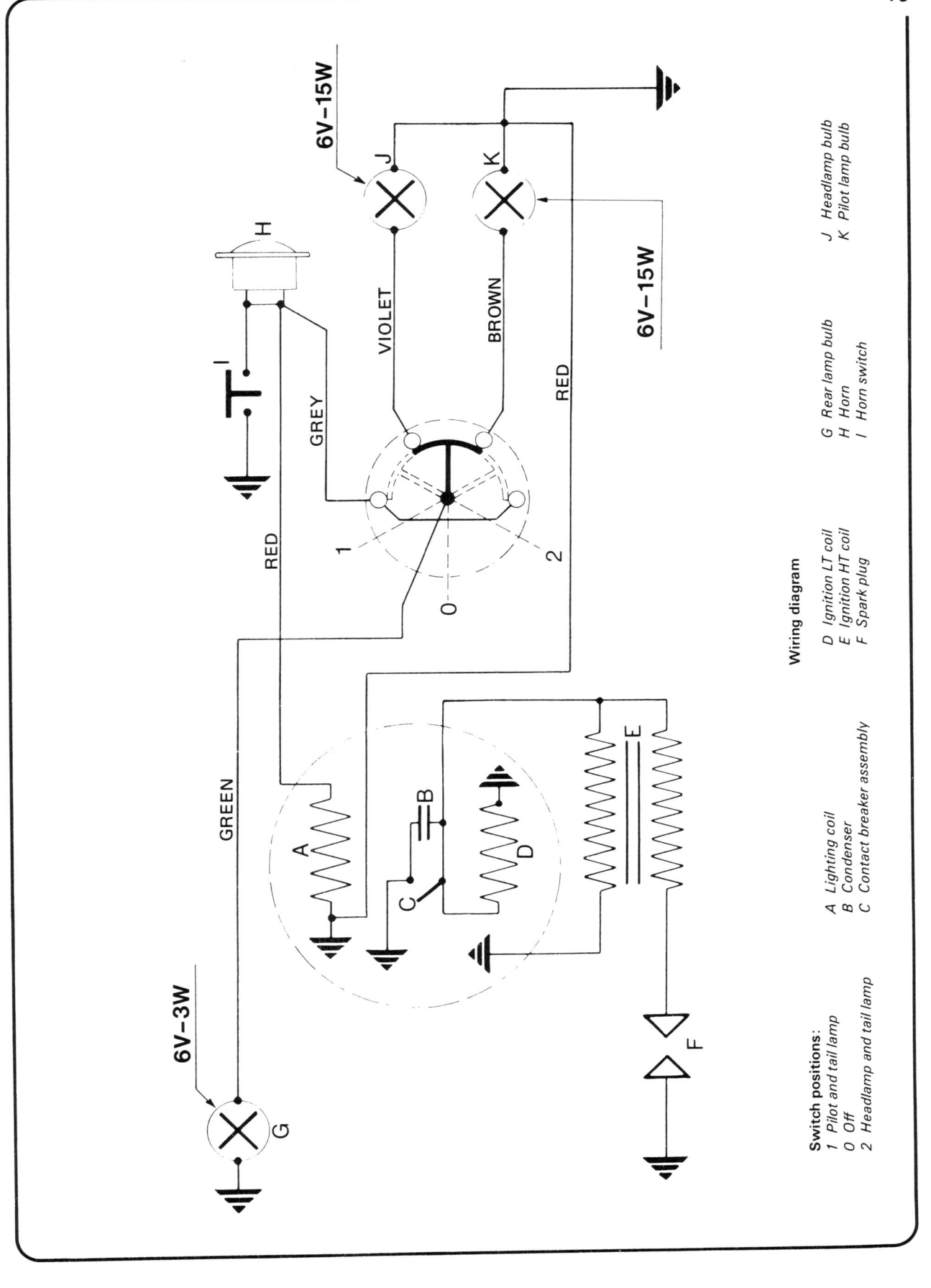

A Lighting coil	G Rear lamp bulb
B Condenser	H Horn
C Contact breaker assembly	I Horn switch
D Ignition LT coil	J Headlamp bulb
E Ignition HT coil	K Pilot lamp bulb
F Spark plug	

Switch positions:
1 Pilot and tail lamp
0 Off
2 Headlamp and tail lamp

Metric conversion tables

Inches	Decimals	Millimetres		Millimetres to Inches			Inches to Millimetres
			mm	Inches		Inches	mm
1/64	0.015625	0.3969	0.01	0.00039		0.001	0.0254
1/32	0.03125	0.7937	0.02	0.00079		0.002	0.0508
3/64	0.046875	1.1906	0.03	0.00118		0.003	0.0762
1/16	0.0625	1.5875	0.04	0.00157		0.004	0.1016
5/64	0.078125	1.9844	0.05	0.00197		0.005	0.1270
3/32	0.09375	2.3812	0.06	0.00236		0.006	0.1524
7/64	0.109375	2.7781	0.07	0.00276		0.007	0.1778
1/8	0.125	3.1750	0.08	0.00315		0.008	0.2032
9/64	0.140625	3.5719	0.09	0.00354		0.009	0.2286
5/32	0.15625	3.9687	0.1	0.00394		0.01	0.254
11/64	0.171875	4.3656	0.2	0.00787		0.02	0.508
3/16	0.1875	4.7625	0.3	0.01181		0.03	0.762
13/64	0.203125	5.1594	0.4	0.01575		0.04	1.016
7/32	0.21875	5.5562	0.5	0.01969		0.05	1.270
15/64	0.234375	5.9531	0.6	0.02362		0.06	1.524
1/4	0.25	6.3500	0.7	0.02756		0.07	1.778
17/64	0.265625	6.7469	0.8	0.03150		0.08	2.032
9/32	0.28125	7.1437	0.9	0.03543		0.09	2.286
19/64	0.296875	7.5406	1	0.03937		0.1	2.54
5/16	0.3125	7.9375	2	0.07874		0.2	5.08
21/64	0.328125	8.3344	3	0.11811		0.3	7.62
11/32	0.34375	8.7312	4	0.15748		0.4	10.16
23/64	0.359375	9.1281	5	0.19685		0.5	12.70
3/8	0.375	9.5250	6	0.23622		0.6	15.24
25/64	0.390625	9.9219	7	0.27559		0.7	17.78
13/32	0.40625	10.3187	8	0.31496		0.8	20.32
27/64	0.421875	10.7156	9	0.35433		0.9	22.86
7/16	0.4375	11.1125	10	0.39370		1	25.4
29/64	0.453125	11.5094	11	0.43307		2	50.8
15/32	0.46875	11.9062	12	0.47244		3	76.2
31/64	0.48375	12.3031	13	0.51181		4	101.6
1/2	0.5	12.7000	14	0.55118		5	127.0
33/64	0.515625	13.0969	15	0.59055		6	152.4
17/32	0.53125	13.4937	16	0.62992		7	177.8
35/64	0.546875	13.8906	17	0.66929		8	203.2
9/16	0.5625	14.2875	18	0.70866		9	228.6
37/64	0.578125	14.6844	19	0.74803		10	254.0
19/32	0.59375	15.0812	20	0.78740		11	279.4
39/64	0.609375	15.4781	21	0.82677		12	304.8
5/8	0.625	15.8750	22	0.86614		13	330.2
41/64	0.640625	16.2719	23	0.90551		14	355.6
21/32	0.65625	16.6687	24	0.94488		15	381.0
43/64	0.671875	17.0656	25	0.98425		16	406.4
11/16	0.6875	17.4625	26	1.02362		17	431.8
45/64	0.703125	17.8594	27	1.06299		18	457.2
23/32	0.71875	18.2562	28	1.10236		19	482.6
47/64	0.734375	18.6531	29	1.14173		20	508.0
3/4	0.75	19.0500	30	1.18110		21	533.4
49/64	0.765625	19.4469	31	1.22047		22	558.8
25/32	0.78125	19.8437	32	1.25984		23	584.2
51/64	0.796875	20.2406	33	1.29921		24	609.6
13/16	0.8125	20.6375	34	1.33858		25	635.0
53/64	0.828125	21.0344	35	1.37795		26	660.4
27/32	0.84375	21.4312	36	1.41732		27	685.8
55/64	0.859375	21.8281	37	1.4567		28	711.2
7/8	0.875	22.2250	38	1.4961		29	736.6
57/64	0.890625	22.6219	39	1.5354		30	762.0
29/32	0.90625	23.0187	40	1.5748		31	787.4
59/64	0.921875	23.4156	41	1.6142		32	812.8
15/16	0.9375	23.8125	42	1.6535		33	838.2
61/64	0.953125	24.2094	43	1.6929		34	863.6
31/32	0.96875	24.6062	44	1.7323		35	889.0
63/64	0.984375	25.0031	45	1.7717		36	914.4

Metric conversion tables

1   Imperial gallon = 8 Imp pints = 1.16 US gallons = 277.42 cu in = 4.5459 litres

1   US gallon = 4 US quarts = 0.862 Imp gallon = 231 cu in = 3.785 litres

1   Litre = 0.2199 Imp gallon = 0.2642 US gallon = 61.0253 cu in = 1000 cc

Miles to Kilometres		Kilometres to Miles	
1	1.61	1	0.62
2	3.22	2	1.24
3	4.83	3	1.86
4	6.44	4	2.49
5	8.05	5	3.11
6	9.66	6	3.73
7	11.27	7	4.35
8	12.88	8	4.97
9	14.48	9	5.59
10	16.09	10	6.21
20	32.19	20	12.43
30	48.28	30	18.64
40	64.37	40	24.85
50	80.47	50	31.07
60	96.56	60	37.28
70	112.65	70	43.50
80	128.75	80	49.71
90	144.84	90	55.92
100	160.93	100	62.14

lb f ft to Kg f m		Kg f m to lb f ft		lb f/in^2: Kg f/cm^2		Kg f/cm^2: lb f/in^2	
1	0.138	1	7.233	1	0.07	1	14.22
2	0.276	2	14.466	2	0.14	2	28.50
3	0.414	3	21.699	3	0.21	3	42.67
4	0.553	4	28.932	4	0.28	4	56.89
5	0.691	5	36.165	5	0.35	5	71.12
6	0.829	6	43.398	6	0.42	6	85.34
7	0.967	7	50.631	7	0.49	7	99.56
8	1.106	8	57.864	8	0.56	8	113.79
9	1.244	9	65.097	9	0.63	9	128.00
10	1.382	10	72.330	10	0.70	10	142.23
20	2.765	20	144.660	20	1.41	20	284.47
30	4.147	30	216.990	30	2.11	30	426.70

Index

A
Acknowledgements 2
About this manual 2
Adjustments:
    Brakes – front and rear 65
    Carburettor 9, 44
    Contact breaker 9, 47
    Drivebelt tension 9
    Handlebars 11
    Headlamp beam height 70
    Saddle 55
    Tyre pressures 8, 61
Air cleaner 44
Automatic:
    Clutch unit – variable ratio models 36
    Speed governor 35

B
Bearings:
    Big end 14
    Main 14, 18
    Small end 14
    Steering head 4
    Wheel:
        Front 62
        Rear 65
Brakes:
    Front 62
    Rear 65, 67
Bulbs – replacement
    General comments 11
    Headlamps 70
    Tail lamp 70
Buying:
    Spare parts 7
    Tools 13

C
Carburettors:
    Adjustment 44
    Dismantling, examination 40
    Fault diagnosis 45
    Removal 40
    Specifications 39

Cables:
    Brake – front and rear 11
    Decompressor 11
    Lubrication 10
    Speedometer 59
    Throttle 11
Chapter contents:
    1.1 Engine 14
    1.2 Transmission 27
    2 Fuel system and lubrication 39
    3 Ignition system 46
    4 Frame and forks 49
    5 Wheels, brakes and tyres 61
    6 Electrical system 69
Checks:
    Coil ignition 47
    Contact breaker points 9, 46
    Electrical system 8, 69
    Ignition timing 46
    Oil level:
        Front forks 9
        Rear hub 9
    Spark plug gap setting 8, 48
    Tyre pressures 8, 61
    Wheel spokes 61
Cleaning:
    Air cleaner 44
    Exhaust system 8, 44
    The machine 59
Clutch:
    Single speed unit – removal, examination and renovation 28
    Automatic unit variable ratio models – removal, examination and models 36
    Specifications 27
Chain – pedal and freewheel unit 35
Coil ignition 47
Condenser 47
Contact breaker 47
Crankcase halves:
    Joining 22
    Separating 18
Crankshaft 18
Cylinder barrel 22

Index

Cylinder head 22

D

Decarbonising 9, 22
Description – general:
    Electrical system 69
    Engine 14
    Frame and forks 49
    Fuel system 39
    Ignition system 46
    Transmission 27
    Wheels, brakes and tyres 61
Dimensions and weights 6
Drivebelt adjustment 37
Dust caps and tyre valves 68

E

Electrical system:
    Fault diagnosis 72
    Flywheel magneto/generator unit 69
    Headlamp 70
    Horn 70
    Specifications 69
    Tail lamp 71
    Switches 70
    Wiring diagram 73
    Wiring layout and examination 71
Engine:
    Crankshaft – examination and replacement 18
    Crankcase halves:
        Joining 22
        Separating 18
    Cylinder barrel 22, 24
    Cylinder head 22
    Decarbonising 9, 19, 22
    Dismantling – general 16
    Fault diagnosis 38
    Gudgeon pin 23
    Lubricating 39
    Piston and rings 19, 23
    Reassembly 22
    Refitting in frame 25
    Removal 15
    Specification 14
    Starting and running a rebuilt unit 25
    Valve grinding 22
Exhaust cleaning 8

F

Fault diagnosis:
    Clutch 38
    Electrical system 72
    Engine 38
    Frame and forks 60
    Fuel system 45
    Ignition system 48
    Lubrication system 45
    Transmission 38
    Wheels, brakes and tyres 68
Filter – air 44
Frame and forks:
    Centre stand 53
    Fault diagnosis 60
    Frame – examination and renovation 53
    Front forks – rigid, removal, examination and renewal 49
    Front forks – hydraulic, removal & reassembly 51
    Rear suspension unit 53
    Saddle 55
    Swinging arm rear suspension 53
    Freewheel unit and pedal chain 35
    Front wheel 62
    Fuel system:

Carburettor:
    Adjustment 9, 44
    Dismantling and examination 40
    Removal 40
Air cleaner 44
Fault diagnosis 45
Petrol load pipe 40
Petrol tank 39
Petrol cap 40

G

Gudgeon pin 23

H

Handlebar switch – horn & lights 70
Headlamp – beam height adjustment 70
High tension leads – spark plug 48
Horn – location 71

I

Ignition system:
    Coil – removal and replacement 47
    Condenser 47
    Contact breaker 47
    Fault diagnosis 48
    Spark plug:
        Checking and setting gap 48
        Colour chart:
            Operating conditions 43
    Specifications 46
    Timing 46

L

Legal obligations 8, 68
Lubrication system 39
    Control cables 10
    General 10
    Leading link front suspension 10
    Speedometer cable 49
    Wheel bearings 62
Lubricants, recommended 12

M

Maintenance, adjustments and capacities 12
Maintenance – routine 8–11
Metric conversion tables 74–75

O

Ordering:
    Spare parts 7
    Tools 13

P

Pedal chain and freewheel unit 35
Petrol feed pipe 40
Petrol tank 39
Petrol cap 40
Piston and rings 19

R

Rear suspension unit 53
Rear wheel 62, 65
Recommended lubricants 12
Repair information 11
Rings and piston 19
Routine maintenance 11

S

Saddle 55
Spark plug:
    Checking and setting gap 48
    Colour chart – operating conditions 43

Index

Specifications:
    Bulbs 69
    Electrical system 69
    Engine 14
    Frame and forks 49
    Fuel system 39
    Ignition system 47
    Lubrication 39
    Transmission 27
    Wheels, brakes and tyres 61
Speedometer head and drive 59
Statutory requirements 8, 68
Suspension units – rear 53
Swinging arm rear suspension 53
Switch – handlebar 70

T

Tools 13
Transmission:
    Automatic speed governor 35
    Chainwheel, pedals and shaft 37
    Clutch:
        Single speed unit – removal, examination and renovation 36
        Automatic – variable ratio – removal and renovation 36
    Drivebelt – adjustment 37
    Fault diagnosis 38
    Pedals, chain and freewheel unit 35
    Rear hub unit – single speed models 30
    Rear hub unit – variable ratio models 32
    Rear pulley – single speed models 28
    Specifications 27
Tyres:
    Pressure 8, 61
    Removal and replacement 68
    Colour instructions 63
    Valves and dust caps 68

W

Weights and dimensions 6
Wheel bearings 62–65
Wheels:
    Front 62
    Rear 62–65
Wiring diagram 73
Wiring layout and examination 71
Working conditions and tools 13

Printed by
Haynes Publishing Group
Sparkford Yeovil Somerset
England